UN-NATURAL DISASTERS

IOWA'S EF5 TORNADO AND THE HISTORIC FLOODS OF 2008

BY **TERRY SWAILS** AND **CAROLYN S. WETTSTONE**

*To the Cordova Library
Best Wishes!
Terry Swails*

*All the Best!
Carolyn S. Wettstone*

DISCARD

ACKNOWLEDGMENTS

To all those who gave of their time, talents, and thoughts during this period of major upheaval and crisis, we are extremely grateful. Their efforts underscore the commitment and dedication of city administrators, state and city engineers, and law enforcement officials who graciously shuffled chaotic schedules to help document these historic events. We also thank the victims of these disasters who opened up their hearts, bared their souls, and helped us understand and appreciate the immense physical and emotional toll these tragedies have wrought. We are especially grateful to Ruth Hahn of Parkersburg, Iowa, for her tireless assistance; Brenda Brock, meteorologist for the National Weather Service in Des Moines, Iowa, for her expertise and perspective; and Shirley Machonis, designer at Farcountry Press, for her incredible talents and insight. Many thanks also to the National Weather Service, especially the committed staff of the Des Moines and Quad City offices, as well as the Storm Prediction Center, the National Oceanic and Atmospheric Administration, the U.S. Corps of Engineers, and the U.S. Geological Survey. To those who captured and shared the powerful images in this book, your work will serve as a permanent reminder of the explosive atmospheric power that was unleashed on Iowa during the spring of 2008. We commend your efforts. Lastly, to the people of Iowa, and the volunteers from around the world, your spirit, will, and determination shines throughout Iowa's darkest hours.

This book is dedicated to our daughter Eden Malone and to our parents, Arnold (Boo) and Rose Swails, and Dorathy D. Wettstone and in memoriam to Dr. Robert E. Wettstone.

ISBN 13: 978-1-59152-054-2
ISBN 10: 1-59152-054-1

© 2008 Terry Swails and Carolyn S. Wettstone
Text © Terry Swails and Carolyn S. Wettstone

FRONT COVER: Tornado, COURTESY KYLE AND TIM WOLTHOFF;
flood images left to right, COURTESY TRAVIS MUELLER, UNIVERSITY OF IOWA, UNIVERSITY OF IOWA, UNIVERSITY OF IOWA.
BACK COVER: Flood, COURTESY DAVID GREEDY, GETTY IMAGES.

Library of Congress Cataloging-in-Publication data - on file

This book may not be reproduced in whole or in part by any means (with the exception of short quotes for the purpose of review) without the permission of the publisher.

Produced by Sweetgrass Books, P.O. Box 5630, Helena, MT 59604
(800) 821-3874

Created, produced, and designed in the United States.
Printed in Canada.

14   13   12   11   10   09   08   1   2   3   4   5   6   7

A LETTER FROM IOWA GOVERNOR CHET CULVER

Since our severe weather began on May 25th, Iowans have been faced with the worst natural disaster in our history. This "500-year flood" displaced 40,000 Iowans and disrupted the lives of countless others.

Iowa farmers, who feed and fuel the world, saw their fields washed away. Businesses large and small faced, and continue to face, immense financial challenges. But I believe we're at our best with our backs up against the wall. In the face of great difficulties, Iowans "locked arms" and came together to provide a calm within the storm.

For every Iowan who has been, and remains, displaced by the tornadoes and floods, there are so many citizens who rose to the occasion to provide comfort and care to others. I have seen up close the bravery of Iowans, young and old, statewide, who, when confronted with hardship, kept their optimism, rolled up their sleeves, and went to work to help their fellow Iowans. And I believe that's a story of the tornadoes and floods of 2008 that should always be remembered.

Four thousand brave soldiers and airmen of the Iowa National Guard worked around the clock to fill sandbags, build up levees, and keep communities safe–the largest deployment of our "hometown heroes" since the Civil War. Thousands of state, county, and municipal employees worked overtime. And more than 50,000 Iowans volunteered, donating their time and their labor to their neighbors to help.

And when I think of "Iowa heroes," I'm reminded of the young Boy Scouts at Camp Little Sioux in Monona County. In the face of immense tragedy, they responded to the tornado of June 11th with heroism and bravery. They showed all of us what "courage" truly means.

The tornadoes and the floods were indiscriminate in the pain they caused. But I believe there is another part of the story we must continue to tell: that of the goodness in the hearts of Iowans.

Traveling all over the state in the aftermath of the storms, I saw that the first responders did not distinguish between old and young. The Guardsmen and women did not ask if you were black or white, urban or rural. The volunteers clearing debris and sandbagging did not ask if you were gay or straight, poor or rich. Instead, what they asked was, "How can we help?"

This is the spirit of "One Iowa, with One Unlimited Future!" Because we are "One Iowa," we will rebuild together. We will rebuild, with faith in each other, and in our future.

Chet Culver

CREDITS LEFT TO RIGHT: SHIRLEY A. MILLER, SCOTT TJABRING, NATIONAL WEATHER SERVICE, NATIONAL WEATHER SERVICE, NATIONAL WEATHER SERVICE

GENESIS of a GIANT

At 3:30 P.M. on Sunday, May 25, lead forecaster Jack Hales of the Storm Prediction Center in Norman, Oklahoma, had seen enough. He tapped the send key on the alert he had prepared, and seconds later it shimmered on the monitors of weather offices around the country, tripping bells and alarms. The alert was designated Tornado Watch #363, reflecting the number of tornado watches already issued in 2008. This particular watch covered much of Iowa, including Butler County and the towns of Parkersburg and New Hartford. As forecasters and media outlets digested the contents of the text, it quickly became clear this was not your average watch. In bold red letters a sentence glared from the page: **"This is a particularly dangerous situation."**

**URGENT - IMMEDIATE BROADCAST REQUESTED
TORNADO WATCH NUMBER 363
NWS STORM PREDICTION CENTER NORMAN OK
330 PM CDT SUN MAY 25 2008**

THE NWS STORM PREDICTION CENTER HAS ISSUED A
TORNADO WATCH FOR PORTIONS OF

A LARGE PART OF IOWA

EFFECTIVE THIS SUNDAY AFTERNOON AND EVENING FROM
330 PM UNTIL 1000 PM CDT.

...THIS IS A PARTICULARLY DANGEROUS SITUATION...

DESTRUCTIVE TORNADOES...LARGE HAIL TO 3 INCHES IN
DIAMETER... THUNDERSTORM WIND GUSTS TO 80 MPH...AND
DANGEROUS LIGHTNING ARE POSSIBLE IN THESE AREAS.

THE TORNADO WATCH AREA IS APPROXIMATELY ALONG AND
75 STATUTE MILES NORTH AND SOUTH OF A LINE FROM 60
MILES SOUTHWEST OF FORT DODGE IOWA TO 50 MILES EAST
SOUTHEAST OF CEDAR RAPIDS IOWA. FOR A COMPLETE
DEPICTION OF THE WATCH SEE THE ASSOCIATED WATCH
OUTLINE UPDATE (WOUS64 KWNS WOU3).

DISCUSSION...VERY TO EXTREMELY UNSTABLE AIR MASS
ACROSS MUCH OF IA AHEAD OF COLD FRONT. THUNDER-
STORMS WILL DEVELOP RAPIDLY AND WITH STRONG DEEP
LAYER SHEAR QUICKLY BECOME SUPERCELLS. VERY LARGE
HAIL AND TORNADOES ARE LIKELY WITH ANY SUPERCELL.
POTENTIAL FOR LONG LIVED SUPER CELLS WITH STRONG
TORNADOES.
AVIATION...TORNADOES AND A FEW SEVERE THUNDERSTORMS
WITH HAIL SURFACE AND ALOFT TO 3 INCHES. EXTREME
TURBULENCE AND SURFACE WIND GUSTS TO 70 KNOTS.
A FEW CUMULONIMBI WITH MAXIMUM TOPS TO 550. MEAN
STORM MOTION VECTOR 24035.

COURTESY NWS STORM PREDICTION CENTER

PDS WATCH (particularly dangerous situation) issued for Butler County at 3:30 p.m.

Tornadoes	
Probability of 2 or more tornadoes	High (80%)
Probability of 1 or more strong (F2–F5) tornadoes	Mod (50%)
Wind	
Probability of 10 or more severe wind events	High (70%)
Probability of 1 or more wind events > 65 knots	Mod (60%)
Hail	
Probability of 10 or more severe hail events	High (70%)
Probability of 1 or more hailstones > 2 inches	Mod (50%)
Combined Severe Hail/Wind	
Probability of 6 or more combined severe hail/wind events	High (90%)

Note the high potential assessment and the 50 percent risk of at least an EF2 tornado. COURTESY NOAA/NWS/STORM PREDICTION CENTER

The potential existed for one or more long-lived destructive tornadoes. Nothing closer to the truth could have been prophesied for the unsuspecting town of Parkersburg, Iowa.

Despite the strong wording, life went on as usual in the town of Parkersburg, population 1,900. After a long and snowy winter, spring had proved to be unusually cool and damp. In the rich farming fields Iowa is known for, corn was just beginning to show. On this Sunday afternoon, the sun was out and temperatures flirted with 80 degrees F. And for the first time all year it had grown noticeably muggy, more typical of May. It was also Memorial Day weekend, and the bright warm weather was especially welcome. Despite the speculation of potentially bad storms, residents were enjoying the holiday. Who wouldn't? After all, threats of tornadoes were an annual rite of spring, and, while Iowans respected twisters, the ones that occasionally visited were typically small and confined to the open fields.

At the National Weather Service in Des Moines, Warning Coordinator Jeff Johnson had a feeling he was in for a rough day of severe weather. By noon the computer models and morning's soundings had triggered enough concern for him to call emergency management coordinators around central Iowa. By 1:30 P.M. Johnson was so convinced of serious storms he called in his warning team early. He also took a few minutes out of the busy afternoon to be interviewed by a local television station about tornado safety in cars.

Karl Jungbluth, the science and operations officer at the Des Moines

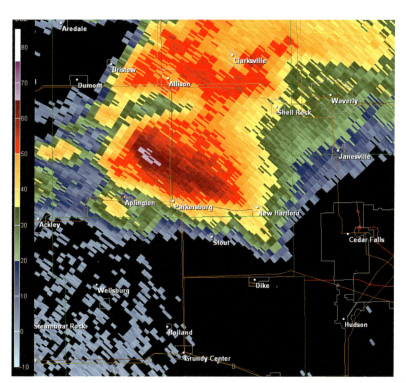 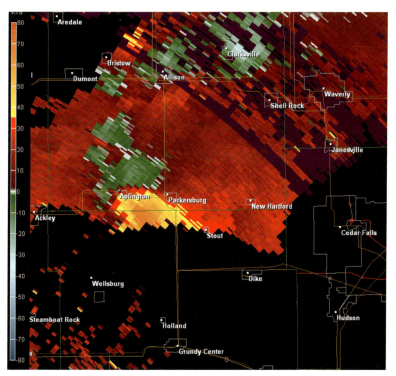

4:50 P.M. Base reflectivity and velocity image from the NWS WSR-88D in Des Moines, two minutes after initial tornado touchdown on the Butler-Grundy county line south of Aplington. Shades of green depict motion toward radar, eighty miles to the southwest, whereas reds and oranges depict motion away from the radar. COURTESY NATIONAL WEATHER SERVICE DSM

**BULLETIN - EAS ACTIVATION REQUESTED
TORNADO WARNING
NATIONAL WEATHER SERVICE DES MOINES IA
422 PM CDT SUN MAY 25 2008**

THE NATIONAL WEATHER SERVICE IN DES MOINES HAS ISSUED A

✷ TORNADO WARNING FOR...
NORTHWESTERN GRUNDY COUNTY IN CENTRAL IOWA...
NORTHEASTERN HARDIN COUNTY IN CENTRAL IOWA...
BUTLER COUNTY IN NORTH CENTRAL IOWA...
SOUTHEASTERN FRANKLIN COUNTY IN NORTH CENTRAL IOWA...

✷ UNTIL 445 PM CDT.

✷ AT 418 PM CDT...NATIONAL WEATHER SERVICE DOPPLER RADAR INDICATED A SEVERE THUNDERSTORM CAPABLE OF PRODUCING A TORNADO NEAR IOWA FALLS...OR 41 MILES WEST OF WATERLOO...MOVING NORTHEAST AT 39 MPH.

✷ THE TORNADO WILL BE NEAR...
ACKLEY BY 425 PM CDT...
APLINGTON BY 440 PM CDT...

THIS TORNADO WARNING REPLACES THE SEVERE THUNDERSTORM WARNING THAT WAS IN EFFECT FOR THE SAME AREA. GO TO A BASEMENT OR SMALL INTERIOR ROOM ON THE LOWEST FLOOR!

COURTESY NWS STORM PREDICTION CENTER

> Outside of the United States, many different names are used to describe tornadoes. Here are a few: Japan, "tatsu-maki"; Germany, "trombe"; Holland, "wervel storm"; Italy, "tramba d'aria"; China, "lung geng fong"; and Australia, "tornado" or "cock-eyed bob."

office, arrived about this time; he recalled his thoughts: "We had time to analyze things, which is the plan when anticipating significant weather. Oftentimes we get severe weather environments, but we usually see negatives . . . but we really couldn't find any negatives, the parameters were all positives for severe weather. We were all concerned as the afternoon went on."

Back in Parkersburg, weather conditions were rapidly deteriorating. In the time it takes to cook a couple of burgers on the grill, the skies to the southwest had billowed with clouds that some say had a hazy tinge of yellow and gray. The source of the change was a massive thunderstorm that had rumbled into existence. Feeding on the extreme instability that had been analyzed by the Storm Prediction Center, the storm exploded as it soared through a layer of dry air aloft and potent shear. Rotating in the distance all by itself was what meteorologists call a mesocyclone, or supercell, an elite and dangerous brand of thunderstorm known to produce violent weather. With nothing to inhibit its growth, the mushrooming storm gobbled up energy as it climbed to heights eight miles above the ground.

The storm's evolution was being closely monitored eighty miles away at the National Weather Service in Des Moines. Johnson and his staff were focused on a portion of the thunderstorm where winds near the ground were blowing in opposite directions. This portion of the storm, indicated by different colors on the powerful WSR-88D Doppler, was a strong signal that a tornado was developing. At 4:22 P.M. the telltale circulation known as a couplet accelerated, and Johnson quickly put out an urgent warning that a tornado was forming near Ackley and would be approaching Butler County. The city of Parkersburg was in the path of the storm.

Predicting something as fickle as a supercell thunderstorm is one of forecasting's biggest challenges. Evolving by the second, these complex storms are in a continual state of flux as they ebb and flow. As it turns out, the supercell Johnson warned about did not produce the Parkersburg tornado—a new storm had developed on its flank a few miles south. Flourishing in a more favorable environment, it was strangling the initial storm and absorbing its energy. Johnson recalled having to switch strategies: "We started focusing on the southern storm. It actually went from virtually no storm at 4:20 P.M. to producing the initial Parkersburg

**BULLETIN - EAS ACTIVATION REQUESTED
TORNADO WARNING
NATIONAL WEATHER SERVICE DES MOINES IA
446 PM CDT SUN MAY 25 2008**

THE NATIONAL WEATHER SERVICE IN DES MOINES HAS ISSUED A

* TORNADO WARNING FOR...
  NORTHERN GRUNDY COUNTY IN CENTRAL IOWA...
  SOUTHEASTERN BUTLER COUNTY IN NORTH CENTRAL IOWA...

* UNTIL 530 PM CDT.

* AT 441 PM CDT...NATIONAL WEATHER SERVICE DOPPLER RADAR INDICATED A SEVERE THUNDERSTORM CAPABLE OF PRODUCING A TORNADO 7 MILES SOUTHWEST OF APLINGTON...OR 33 MILES WEST OF WATERLOO... MOVING NORTHEAST AT 36 MPH.

* THE TORNADO WILL BE NEAR...
  APLINGTON BY 455 PM CDT...
  PARKERSBURG BY 500 PM CDT...
  ALLISON BY 510 PM CDT...
  SHELL ROCK AND CLARKSVILLE BY 520 PM CDT...

COURTESY NWS STORM PREDICTION CENTER

tornado in thirty minutes. The mesocyclone became the dominant storm within ten minutes of the original tornado warning."

In Parkersburg, the growing threat of serious weather was now apparent. Even though the powerhouse storm was a half hour away, its anvil dominated the southwestern horizon. Clouds streaked across the sky above Parkersburg and a steady wind blew through the streets. Residents noted some apprehension as the storm's presence became apparent. However, despite the tornado warning and the direct mention of Parkersburg as a target, most townspeople continued business as usual. Meanwhile, at the base of the growing storm, a wall cloud emerged and began to slowly rotate.

Coincidentally, Rod Donavon, a senior meteorologist at the Des Moines office, and his family were at a graduation party in his hometown of Clarksville, about twenty minutes from Parkersburg. He was aware of the tornado threat and took off with his camera to see if he could document it. As he drove toward Parkersburg, his parents listened to the chatter on their Butler County scanner and advised him of the storm by phone.

Back in Des Moines, Johnson watched with apprehension as the second storm exploded before his eyes. With each scan of the Doppler, it was evident that the storm's ominous structure and strong rotation was

**SEVERE WEATHER STATEMENT
NATIONAL WEATHER SERVICE DES MOINES IA
451 PM CDT SUN MAY 25 2008**

IAC023-075-252230-
/O.CON.KDMX.TO.W.0013.000000T0000Z-080525T2230Z/
BUTLER IA-GRUNDY IA-
451 PM CDT SUN MAY 25 2008

...A TORNADO WARNING REMAINS IN EFFECT UNTIL 530 PM CDT FOR NORTHERN GRUNDY AND SOUTHERN BUTLER COUNTIES...

AT 450 PM CDT...TRAINED WEATHER SPOTTERS REPORTED A TORNADO. THIS TORNADO WAS LOCATED NEAR APLINGTON...OR 30 MILES WEST OF WATERLOO...MOVING NORTHEAST AT 33 MPH.

THE TORNADO WILL BE NEAR...
  PARKERSBURG BY 500 PM CDT...
  8 MILES NORTHWEST OF STOUT BY 505 PM CDT...
  8 MILES SOUTHEAST OF ALLISON AND 8 MILES NORTHWEST OF NEW HARTFORD
  BY 515 PM CDT...
  SHELL ROCK AND 6 MILES SOUTHEAST OF CLARKSVILLE BY 525 PM CDT...

AT 448 PM...SPOTTERS REPORTED A CONFIRMED TORNADO AT THE INTERSECTION OF COUNTY ROAD T19 AND HIGHWAY 57 IN BUTLER COUNTY...HEADING TOWARD APPLINGTON.

A TORNADO WATCH REMAINS IN EFFECT UNTIL 900 PM CDT SUNDAY EVENING FOR NORTHWESTERN IOWA.

COURTESY NWS STORM PREDICTION CENTER

Tornado entering Parkersburg at 4:59 p.m. COURTESY MILES HUMPHREY

increasing. Johnson issued a follow-up warning at 4:46 p.m. that a tornado was due in Parkersburg by 5:00 p.m.

The storm advanced at a steady, relentless pace, grinding along south of the small town of Aplington.

In Des Moines, spotter reports started flooding in. The storm had finally put down a confirmed tornado at 4:48 p.m. on the Butler-Grundy county line. The sobering reality of a twister on the ground sent a heightened wave of adrenalin and concern throughout the warning staff. Jungbluth recalled, "Once we heard we had a real tornado, we were all just very conscious of it. I can remember reminding each other what our specific tasks were. We were all sort of a part of it and we all turned very business-like."

With clear sailing ahead, a healthy and growing tornado took aim at Parkersburg, roughly six miles away.

At 4:51 p.m., with new information of a potentially significant tornado on the ground and evidence of a couplet or hook echo on radar, Johnson issued a follow-up statement saying a tornado would be near

**SEVERE WEATHER STATEMENT
NATIONAL WEATHER SERVICE DES MOINES IA
458 PM CDT SUN MAY 25 2008**

BUTLER IA-GRUNDY IA-
458 PM CDT SUN MAY 25 2008

...A TORNADO WARNING REMAINS IN EFFECT UNTIL 530 PM CDT FOR NORTHERN GRUNDY AND SOUTHERN BUTLER COUNTIES...

AT 455 PM CDT...TRAINED WEATHER SPOTTERS REPORTED A TORNADO. THIS TORNADO WAS LOCATED NEAR PARKERSBURG...OR 25 MILES WEST OF WATERLOO... MOVING NORTHEAST AT 40 MPH.

THE TORNADO WILL BE NEAR...
   NEW HARTFORD BY 510 PM CDT...
   SHELL ROCK AND 9 MILES SOUTHEAST OF CLARKSVILLE BY 515 PM CDT...

AT 452...A SPOTTER REPORTED A LARGE TORNADO NEAR THE INTERSECTION OF HIGHWAY D17 AND HIGHWAY 14.

MULTIPLE TORNADOES ARE FORMING EAST OF APLINGTON.

THIS IS AN EXTREMELY DANGEROUS AND LIFE THREATENING SITUATION. THIS STORM IS CAPABLE OF PRODUCING STRONG TO VIOLENT TORNADOES.

COURTESY NWS STORM PREDICTION CENTER

the city of Parkersburg by 5:00 p.m. Johnson later said, "At that point, the storm really showed the classic tornado signature."

Tornado sirens began to wail in Parkersburg; residents realized the situation was serious and time was running out. Electrical power surged, went on and off, then fitfully failed. Many people stole a few seconds to watch the churning mass of clouds approaching from the southwest, a moment of horror that would be etched in their minds forever.

Survival was now the ultimate goal. Terrified, people grabbed loved ones, pets, bibles, and cell phones, and raced for shelter. Where they ended up in the next few minutes determined their fate. People waited in homes and buildings throughout the town. Some cried, many prayed, ears popped, and everyone felt the power and helplessness that comes from a violent storm. As the sky turned pitch black, Parkersburg entered the arms of an EF5 tornado.

From a safe vantage, storm chasers and emergency management spotters watched Parkersburg's nightmare unfold. On the fringe of town, they called in reports that a large and violent tornado called a "wedge" was about to strike. Urgently and at times frantically, they reported on the storm's assault as it thrashed through the last open field before entering town. At the National Weather Service in Des Moines, the incoming reports were just as alarming as the freshly painted images on the Doppler radar. Johnson and his team were getting a look at a once-in-a-lifetime superstorm capable of producing catastrophic damage. The stakes in terms of human life would never be greater.

Realizing the storm was going to continue on beyond Parkersburg, Johnson put out a follow-up statement for New Hartford.

At 4:58 P.M. the birds in Parkersburg stopped singing. It was as if all living things stopped to take one giant breath. In the eerie quiet, leaves that only minutes before had been straining in the wind drooped from the trees. Menacing black clouds filled the sky, spinning like a top. Lightning flickered.

Then the ominous sound of a monstrous freight train filled the air. Houses began to shake and windows rattled, and then it hit. At precisely 4:59 P.M. the eerie silence exploded into a deafening roar. The air seethed with missiles that seconds before were the structures of Parkersburg. First to fly were roof shingles, followed by punched-out windows and peeled-back siding. Then came roofs and splintered two-by-fours, and,

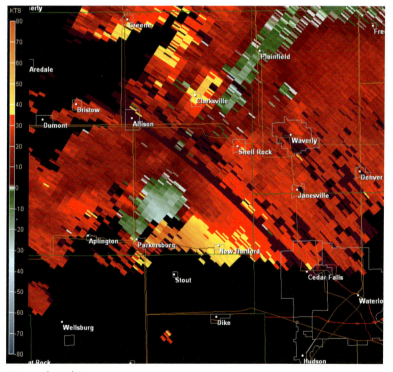

4:59 P.M. The tornado enters Parkersburg (reflectivity left, velocity mode right). COURTESY NATIONAL WEATHER SERVICE DSM

GENESIS OF A GIANT ✸ 11

SEVERE WEATHER STATEMENT
NATIONAL WEATHER SERVICE DES MOINES IA
502 PM CDT SUN MAY 25 2008

BUTLER IA-GRUNDY IA-
502 PM CDT SUN MAY 25 2008

...A TORNADO WARNING REMAINS IN EFFECT UNTIL 530 PM CDT FOR NORTHERN GRUNDY AND SOUTHERN BUTLER COUNTIES...

AT 459 PM CDT...TRAINED WEATHER SPOTTERS REPORTED A TORNADO. THIS TORNADO WAS LOCATED NEAR PARKERSBURG...OR 22 MILES WEST OF WATERLOO... MOVING NORTHEAST AT 33 MPH.

THE TORNADO WILL BE NEAR...
 NEW HARTFORD BY 510 PM CDT...
 SHELL ROCK BY 520 PM CDT...

THIS IS AN EXTREMELY DANGEROUS AND LIFE THREATENING SITUATION. THIS STORM IS CAPABLE OF PRODUCING STRONG TO VIOLENT TORNADOES. IF YOU ARE IN THE PATH OF THIS TORNADO...TAKE COVER IMMEDIATELY!

A WEDGE TORNADO IS EAST OF PARKERSBURG...MOVING EAST.

COURTESY NWS STORM PREDICTION CENTER

> The F5 "Tri-State Tornado" killed 695 people and injured 2,027 on March 18, 1925. The longest-running tornado on record, it was on the ground for 3.5 hours and traveled 219 miles across Missouri, Illinois, and Indiana. At times the storm moved at a forward speed of seventy-three miles per hour. Considered the worst tornado in U.S. history, in its forty-minute journey from Gorham to Parish, Illinois, the tornado killed 541 people and left 1,423 seriously injured. Of the total fatalities, 234 occurred in the town of Murphysboro, Illinois.

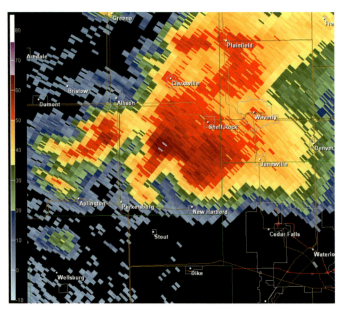

5:04 P.M. Reflectivity vs. velocity (tornado just above the "g" in Parkersburg, or in the area of the Sinclair grain elevator).
COURTESY NATIONAL WEATHER SERVICE DSM

One and a half miles south of New Hartford at 5:05 P.M. COURTESY ROD DONAVON

finally, with house interiors fully exposed, came the contents and history of lifetimes. So fierce were the 205 miles per hour winds that mature trees were stripped of bark; cars and bricks were tossed like Tinkertoys.

The destruction of the southern third of Parkersburg took less than four minutes. This section of town simply ceased to exist.

Traveling from west to east along Highway 57, the half- to three-quarter-mile-wide tornado first leveled the convenience store and a restaurant on the southwest side. Then, one by one, went the grocery store, car wash, bank, a law firm, the high school . . . and 260 Parkersburg homes.

The tornado surged toward New Hartford, and the people of Parkersburg emerged from the rubble. Dazed, confused, and battered, they were shocked to see block after block of complete and utter devastation. Shouts and cries filled the air as residents searched for family, friends, and neighbors. Live power lines, gas leaks, and debris-choked streets added to the chaos. The trapped and injured waited for help; word began to spread among the living that death was another by-product of the storm.

Approaching from the east, Rod Donavon was closing in on the storm. Relying on his NOAA weather radio, his parents' scanner, and the angry twirling sky to the northwest, he knew that what he was looking for was not far away. About five miles southeast of New Hartford, he came face to face with the tornado and gave this firsthand account: "When we first came up over the hill and saw the tornado, it was just a huge mesocyclone reaching for the ground. It had a huge wall cloud, and, initially, I watched it more as a tourist. I hadn't gone out and chased a whole lot because I am usually working storms, and I certainly had never seen anything near that big. Wow, the wall cloud itself was two miles wide. I couldn't tell how wide it was on the ground, but I was in awe at the time I saw it."

SEVERE WEATHER STATEMENT
NATIONAL WEATHER SERVICE DES MOINES IA
511 PM CDT SUN MAY 25 2008

BUTLER IA-GRUNDY IA-
511 PM CDT SUN MAY 25 2008

...A TORNADO WARNING REMAINS IN EFFECT UNTIL 530 PM CDT FOR NORTHERN GRUNDY AND SOUTHERN BUTLER COUNTIES...

AT 508 PM CDT...TRAINED WEATHER SPOTTERS REPORTED A TORNADO. THIS TORNADO WAS LOCATED NEAR NEW HARTFORD...OR 17 MILES WEST OF WATERLOO...MOVING EAST AT 35 MPH.

THE TORNADO WILL BE NEAR...
7 MILES SOUTH OF SHELL ROCK BY 520 PM CDT...

THIS IS AN EXTREMELY DANGEROUS AND LIFE THREATENING SITUATION. THIS STORM IS CAPABLE OF PRODUCING STRONG TO VIOLENT TORNADOES. IF YOU ARE IN THE PATH OF THIS TORNADO...TAKE COVER IMMEDIATELY!

COURTESY NWS STORM PREDICTION CENTER

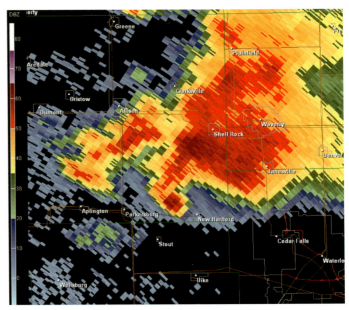

As New Hartford braced for the storm, the small town of 660 residents could see the twister bearing down from the west. For what seemed like a lifetime, it appeared the tornado's path would take it into town. Donavon, now just south of New Hartford, pulled over on Highway 57. He continued his account: "I remember tremendous inflow with winds of forty to fifty miles an hour blowing into the tornado, kicking up dust and making it difficult to see. The tornado was about a half mile to our north, and I definitely didn't want to get any closer. As it moved toward New Hartford, I could see glimpses of the water tower and see everything circling around it. It was interesting, before it got there, the people of New Hartford must have realized what had happened in Parkersburg because there were ten to fifteen cars racing out of town just seconds before the tornado—taking up both lanes heading out of New Hartford."

By now, Donavon was communicating with his co-workers at the

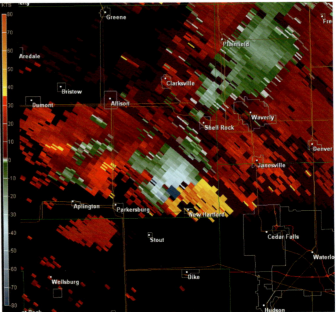

Twister fast approaching New Hartford.
COURTESY NATIONAL WEATHER SERVICE DSM

14 ✦ UN-NATURAL DISASTERS

MESOSCALE DISCUSSION 1029
NWS STORM PREDICTION CENTER NORMAN OK
0508 PM CDT SUN MAY 25 2008

AREAS AFFECTED...IOWA
CONCERNING...TORNADO WATCH 363...
VALID 252208Z - 252315Z

THE SEVERE WEATHER THREAT FOR TORNADO WATCH 363 CONTINUES.

STORMS CAPABLE OF PRODUCING LARGE HAIL...DMGG WINDS...AND TORNADOES WILL CONTINUE TO DEVELOP ACROSS IOWA THIS EVENING.

STORM CLUSTER LOCATED IN NERN IOWA NEAR KCCY APPEARS TO HAVE OVERCOME CIN ON 18Z DVN SOUNDING...AS LOW LEVEL ROTATION IS EVIDENT ON RADAR. DESPITE VEERING OF SURFACE WINDS SEEN ON LATEST 2150Z

OBSERVATIONS AND STORM INTERFERENCE...THESE STORMS MAY STILL HAVE POTENTIAL TO PRODUCE TORNADOES...ESPECIALLY AS THEY MOVE TOWARDS THE WARM FRONT. SOUTHERN END STORM CURRENTLY LOCATED IN BUTLER COUNTY LIKELY HAS BEST TORNADIC POTENTIAL AS IT DRAWS IN HIGHER THETA-E AIR WITH LITTLE INTERFERENCE FROM STORMS TO THE N.

TO THE W...ADDITIONAL THUNDERSTORMS ARE DEVELOPING BEHIND THIS CLUSTER ALONG WSWLY TO ENELY ORIENTED LINE OF CONVERGENCE. STORMS WILL LIKELY RAPIDLY INTENSIFY OVER THE NEXT HOUR.

Tornado approaching New Hartford with 205 mile per hour winds.
COURTESY LAURIE DEGROOTE

Eerie perspective of the tornado. COURTESY MARK RODNEY

National Weather Service in Des Moines, giving his perspective on the storm. At times, debris such as mud and pieces of paper fell on his car. He had planned on going to Parkersburg to call in a damage estimate, but decided against it. In retrospect, he recalled, "I just calmly watched it. It really didn't set in at the time how big it was."

New Hartford narrowly escaped the brunt of the twister; it skimmed the north edge of town. However, the Beaver Hills subdivision and rural areas less than three-quarters of a mile north were not as lucky when the storm roared through at 5:09 P.M.

A notable observer from his farmstead one mile south of New Hartford was U.S. senator Charles Grassley, who watched the tornado still packing winds of 166 to 205 miles per hour. "I was expecting to see a funnel-shaped whirl that you normally associate with tornadoes," he said, but as I watched, it was just a cloud of dust,

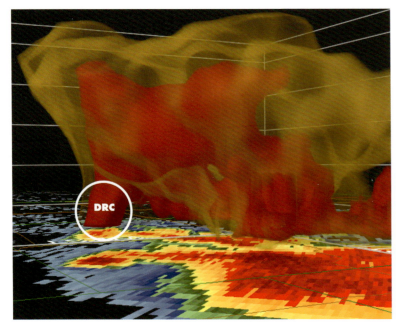

5:08 P.M. KDMX WSR-88D volumetric reflectivity view of the supercell from the east-southeast. This dramatic 3D perspective shows the vertical and horizontal structure of the Parkersburg storm in relation to precipitation indicated at the surface. Precipitation is outlined in blue. Notice the Descending Reflectivity Core (DRC) on the left near the area of the tornado.
COURTESY NATIONAL WEATHER SERVICE DSM

BULLETIN - EAS ACTIVATION REQUESTED
TORNADO WARNING
NATIONAL WEATHER SERVICE DES MOINES IA
521 PM CDT SUN MAY 25 2008

THE NATIONAL WEATHER SERVICE IN DES MOINES HAS ISSUED A

* TORNADO WARNING FOR...
  BLACK HAWK COUNTY IN NORTHEAST IOWA...
  THIS INCLUDES THE CITY OF WATERLOO...
  SOUTHERN BREMER COUNTY IN NORTHEAST IOWA...

* UNTIL 545 PM CDT.

* AT 517 PM CDT...NATIONAL WEATHER SERVICE DOPPLER RADAR WAS TRACKING A LARGE AND EXTREMELY DANGEROUS TORNADO 8 MILES NORTHWEST OF CEDAR FALLS...OR 13 MILES NORTHWEST OF WATERLOO...MOVING EAST AT 34 MPH.

* THE TORNADO WILL BE NEAR...
  CEDAR FALLS AND JANESVILLE BY 530 PM CDT...
  WATERLOO AIRPORT BY 535 PM CDT...
  WATERLOO AND DENVER BY 540 PM CDT...

COURTESY NWS STORM PREDICTION CENTER

wedge-shaped. It didn't have a funnel shape, so I didn't know it was a tornado until I turned on the TV."

At the National Weather Service in Des Moines, reports pouring in grew worse by the minute. Forecasters were now aware they were dealing with a large tornado, and from word out of Parkersburg it might be a killer. Johnson fired off a new tornado warning at 5:21 P.M. for Black Hawk and southern Bremer counties. Of mounting concern was the large metropolitan area of Cedar Falls-Waterloo. A tornado of this magnitude striking such a population base would be a tremendous catastrophe. Johnson remembered his thoughts: "By now we had a pretty itchy trigger finger and we were not going to wait. We had to get out ahead of it, and I started to wonder if the tornado was going to keep making that right turn down the Cedar River valley. I wanted to get Waterloo in the warning, the entire area. I wanted to make sure the players in Waterloo knew they were under a tornado warning."

On a trend slightly north of east, the storm finally crossed the Butler County line, leaving behind a mangled trail of carnage. Reports of the direct hit on Parkersburg alerted hospitals throughout the area to expect casualties. Emergency crews, police, and firefighters from surrounding towns and counties raced to the scene. Upon arrival they were stunned by the spectacle of massive devastation.

By now, spotters had confirmed to the warning team in Des Moines that a large wedge tornado was still on the ground. According to Jungbluth the focus was magnified. "When you hear about a big tornado, you know

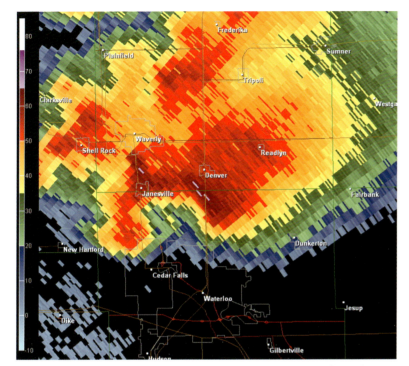 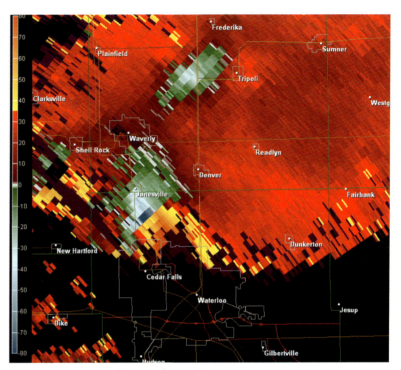

5:26 P.M. Storm relative velocity image from the NWS WSR-88D in Des Moines. The tornado continues to move east just north of Cedar Falls or even right along the northern edge of the town. Shortly after this time, at 5:35 P.M., a 93 mph wind gust was recorded at the Waterloo Municipal Airport, which is in the extreme northwest part of Waterloo. COURTESY NATIONAL WEATHER SERVICE DSM

something really violent can happen, and you get even more serious, more focused," he said. "We are on a professional level, there is no sense of being in the unknown. We know what we have to do and we make sure we get it done, we doubly make sure we get it done."

Johnson added, "There was a high level of excitement in the office, but everyone handled it with a cool head."

Surging into Black Hawk County, the tornado ripped through farmsteads as it traveled east. In less than fifteen minutes, it was on the northern doorstep of Cedar Falls. Residents witnessed the twister crossing U.S. Highway 218 headed north of town. At 5:35 P.M. a wind gust of ninety-three miles per hour was measured at the Waterloo Airport on the extreme northwest edge of the city. While much too close for comfort, the twister narrowly missed the two large metropolitan areas.

Still on a general eastward track, the storm's intensity weakened as it passed just north of Waterloo and skipped across U.S. Highway 63. Despite decreased winds on the order of 135 miles per hour, the tornado was still more than half a mile wide and producing plenty of destruction. Now on its final legs, the storm's track would take it over the small town of Dunkerton, population 750. At 5:37 P.M. Johnson issued a new tornado warning for northern Black Hawk County, including his projection that the twister would be near Dunkerton by 5:50 P.M.

In a final burst of energy, the tornado widened and made a slight turn northeast. As if to show the world how truly incredible it was, it ballooned to a width of 1.2 miles as it flattened homes and farmsteads just

**BULLETIN - EAS ACTIVATION REQUESTED
TORNADO WARNING
NATIONAL WEATHER SERVICE DES MOINES IA
537 PM CDT SUN MAY 25 2008**

THE NATIONAL WEATHER SERVICE IN DES MOINES HAS ISSUED A

* TORNADO WARNING FOR...
  NORTHERN BLACK HAWK COUNTY IN NORTHEAST IOWA...
  THIS INCLUDES THE CITY OF WATERLOO...

* UNTIL 615 PM CDT.

* AT 535 PM CDT...NATIONAL WEATHER SERVICE DOPPLER RADAR WAS TRACKING **A LARGE AND EXTREMELY DANGEROUS TORNADO NEAR WATERLOO AIRPORT...OR NEAR WATERLOO...MOVING EAST AT 35 MPH.**

* THE TORNADO WILL BE NEAR...
  EVANSDALE BY 540 PM CDT...
  RAYMOND BY 545 PM CDT...
  DUNKERTON BY 550 PM CDT...

A CONFIRMED LARGE DAMAGING TORNADO WAS NEAR THE WATERLOO AIRPORT. THIS IS A LIFE THREATENING STORM. EXTENSIVE DAMAGE IS LIKELY.

COURTESY NWS STORM PREDICTION CENTER

The Oklahoma City tornado of May 3, 1999, produced 318 miles per hour winds, the strongest ever measured. A mobile Doppler radar mounted on a truck registered the speed as the funnel moved through south Oklahoma City. The fastest speed previously measured was 286 miles per hour in 1991, in a tornado at Red Rock, Oklahoma.

---

west and north of Dunkerton. Mercifully, it weakened again and finally dissipated in northeast Black Hawk County minutes before moving into the city of Fairbank.

After the tornado lifted, there was a small sigh of relief from the warning team in Des Moines. However welcome, it was brief due to ongoing tornado watches and severe weather over much of Iowa. Johnson recalled, "We were in severe weather mode for hours afterwards. As far as sitting around talking about the tornado, we didn't have time, we were dealing with more weather and we had to get on with our job. The information coming into the office by all indications was really bad. However, we didn't know it was EF5 damage, we had no preconceived notion we were going to take a five." Jungbluth added, "In our job, a tornado is as intense as it gets. When you hear that people are killed, it really sinks in."

Primary region of EF5 damage in Parkersburg.
COURTESY NATIONAL WEATHER SERVICE

Tornado path and equivalent EF damage. COURTESY NATIONAL WEATHER SERVICE DSM

As the news of Parkersburg's nightmare spread from Iowa to the rest of the country, pictures of the battered city showed up in newspapers and television newscasts around the world. Perhaps the May 28 headlines of the *Des Moines Register* summed things up best as it boldly proclaimed, "this was a violent and historic storm." But then again, the people of Parkersburg and New Hartford could have told you that.

Several days after the storm, two National Weather Service survey teams determined the Parkersburg tornado had been on the ground continuously for forty-three miles. In a little over an hour, it destroyed 450 homes; damage estimates ranged up to $90 million. Its maximum wind speed was estimated at 205 miles per hour, making it an EF5 tornado, the strongest possible and the most severe tornado in the eastern half of Iowa in forty years. The storm took the lives of six people in Parkersburg and two in New Hartford. By every conceivable standard, it will be remembered as a storm for the ages, one of the worst tornadoes in the state's history. In essence, it was Iowa's perfect storm. ✹

CONFIRMATION OF DEATH

Brenda Brock is the meteorologist in charge for the National Weather Service in Des Moines. She recounted the impact of the day's tragic events: "That Sunday a squall line had gone through north-central Iowa. One of our warning teams took the northern section and another team took the south part of that line. Jeff Johnson, our warning coordinator, issued the warning for the Parkersburg tornado. Jeff saw that the criteria, all the ingredients that pointed to a large tornado were there. The imagery of the storms reflectivity on the radar was so great that everyone who looked at it said 'that has to be a tornado.' I don't know if this is the right word to use, but the reflectivity was just magnificent. Jeff was very convinced that there was a large tornado so he went ahead and issued the warning.

"I got called by our senior forecaster on duty and I came in. We still didn't have any word about the severity of the tornado; however, the senior forecaster had two unconfirmed deaths so the policy is to call me. We still had no word, which was understandable because the town had been hit so heavily. We had no communication with them. I was here for an hour and we started finding out, one by one, that people had died. I looked around the office once we had a confirmation of death. It was such a solemn atmosphere and it got very quiet here. In all my years of experience, I have never seen such a look on people's faces. It was so somber; oh my god, people died here. The scope of the event was so huge some people's faces were just expressionless."

CREDITS LEFT TO RIGHT: TIM AND KYLE WOLTHOFF, JUSTIN STOCKDALE, DAN BLEVINS, JUSTIN STOCKDALE, APRIL KRAMER

HELL on EARTH

STORIES OF RESCUE AND SURVIVAL

In Parkersburg, New Hartford, and rural areas extending as far east as Dunkerton, everyone has a story to tell about the day the EF5 twister tore into their lives and snatched away people, possessions, and peace of mind. It is impossible to give full justice to the pain, devastation, and agony these survivors suffered; we can give you only a slice of their experience, fully recognizing that until each one of us has walked in their shoes, we can never truly appreciate the terror and tragedy they witnessed.

These are just a few of the stories from that fateful day. What rises to the top of each story is the true instinct of Iowans to help each other and themselves. They will tell you, "I didn't think, I just did it." Unsaid in those simple words were acts of supreme courage and a deep, abiding love for their families and compassion for their neighbors.

A HIGHER CALLING

The golf ball–sized hail started falling while Police Chief Chris Luhring and his family were visiting his wife's parents at their home about five miles north of Parkersburg. "My father-in-law always said that if there is golf ball–sized hail, there is a tornado," said Luhring. "He turned to me and said 'It's time for you to go'."

The police chief and his brother-in-law jumped in a truck and headed toward town. It wasn't long before they could see a massive black cloud advancing on Parkersburg, and their concern for the town and its residents escalated. "I was going 110 miles per hour and my brother-in-law said go faster," said Luhring.

Luhring's roots to Parkersburg are so deep and wide they nearly get tangled. His family was one of the first to settle in the area, so pretty much only the tourists don't know who a Luhring is.

That frightening afternoon, Luhring's focus and fear centered on the people he had grown up with, laughed with, and sat next to in church on Sunday mornings. As he approached town, the sun unexpectedly popped out about two miles north of Parkersburg. "I could see the water tower and strange droppings from the clouds that were very eerie and beautiful at the same time," he said.

Coming into the north side of town, nothing had changed. But when Luhring crested the hill, he was stunned to see that the southern half of his town had "vanished . . . just like it had been wiped off this earth." His parents' farmstead, which had been a reliable part of the landscape, had ceased to exist. "I just assumed my parents were dead. There was nothing I could do about it. I had to move on," he said.

Luhring forced aside his personal feelings and focused on rescue. He had to get to his police vehicle parked at his home. Fighting through the wreckage, he stopped residents along the way, telling them to keep track of who they saw and to write down those names. "I needed to know who was dead and who was alive. We had to keep track of who was there so we could get those names to emergency personnel," he clarified.

In his police vehicle, Luhring headed in the direction of his brother-in-law's home. He didn't get far before he discovered two people injured and trapped in the basement of their home. "A state trooper was lifting up a wall of a house to help get them out. He was risking his life to save

BEFORE: Tomkin residence, 1115 Highway 57, Parkersburg.
COURTESY BUTLER COUNTY ASSESSORS OFFICE

AFTER: Tomkin residence, 1115 Highway 57, Parkersburg.
COURTESY *DES MOINES REGISTER*

them. The couple was barely free when his strength ran out and the wall fell on top of him," he said. The state trooper was hospitalized with serious injuries.

At his brother-in-law's home, Luhring was relieved to find everyone safe. "My brother-in-law was upset and I told him to stop crying and get a pen and paper and write down the names of everyone we know who is safe. I didn't need to waste my time on people who were alive and safe," said Luhring.

His next priority was to locate emergency hot spots and send ambulances. He put his Explorer in four-wheel-drive and launched it over a roof planted right in the middle of the street—just like a kid flying up a bike ramp. He drove across yards, down a bike path, and through ditches to get to the fire station. A command center was set up as well as a temporary morgue. After witnessing the vast destruction, Luhring prepared himself for a large number of fatalities. "I estimated sixty to seventy bodies, others thought a hundred. I set up the morgue on the east side of the fire station, posted a guard, and told him no one was allowed in." Protecting the privacy of those who had died in the storm had become very personal for the police chief. He had just learned that his aunt, seventy-one-year-old Shirley Luhring, had been killed by flying debris. Amazingly, his uncle, Herman, had survived. "They were down in the basement hiding inside a shower. When they found my uncle, he was crimson. Only his eyes were white. The rest of him was covered in blood. He told the first person to arrive that Shirley was gone."

At the command center, Luhring mounted a map of Parkersburg and sectioned off portions of the town. Fire companies were given cans of spray paint, assigned areas to cover, and instructed to go street by street searching homes. "I told them to put a cross on the house if there was a dead person, an 'okay' if there was no one present, and if anyone needed to be saved, that is who you work on."

Butler County's Sheriff Jason Johnson arrived at the command center, and Luhring remembers "putting an orange vest over his head and handing him the sheet with the names of the dead bodies." The two men worked in unison, along with Second Assistant Fire Chief Ryan Siems, running the emergency crews from the command center. "I coordinated with a map and circled the areas where I sent each fire company,"

BEFORE: Ingall residence, 808 N. Johnson, Parkersburg.
COURTESY BUTLER COUNTY ASSESSORS OFFICE

AFTER: Ingall residence, 808 N. Johnson, Parkersburg.
COURTESY *DES MOINES REGISTER*

Luhring said. "The search and recovery went on all night. For hours the firefighter teams kept coming in and heading out. I kept track of how many members there were, where they were from, and where they were going. They kept coming back and saying, 'We didn't find anyone. We didn't find anyone.' I answered, 'This can't be true. This can't be true'."

It wasn't until much later that Luhring saw his sergeant for the first time. "I didn't know until then if he was alive or dead," remembered Luhring. Then came the astounding news that his father and stepmother were alive. "They told me my dad was in the hospital and my brother and mom were alive," he said. After witnessing the destruction of his family's farm, it was a struggle for Luhring to accept the good news. "Even now," Luhring said with a shake of his head, "it is hard to believe. My parents were in the epicenter of the storm. Despite seeing them and knowing they are alive, I sometimes still think they are dead."

Luhring never rested for two days. Finally, in the early morning hours on Wednesday, he lay down in his police rig and caught a two-hour nap. "I have never had an awareness of that type of power. It was totally vertical. I do not normally have the ability to stay up for forty-eight hours—something gave me the strength to do it," he said. "My wife said to me the Tuesday after the tornado, 'Is this the reason you and Jason were born?'" Luhring paused for a moment and reflected. "Maybe it was."

WHEN FIFTY OF YOUR FRIENDS NEED YOU ALL AT ONCE

The call came into New Hartford's fire station shortly after 5:00 P.M.: Parkersburg urgently needed help. A monster tornado had plowed through the southern half of town, and it was a disaster. Volunteer Fire Chief Brad Schipper immediately dispatched a squad of firefighters, along with the town's best ambulance.

Then he jumped into a fire truck with firefighter Gordy Ballhagen and headed to the western edge of New Hartford to staff a lookout point. Schipper turned toward Ballhagen to explain where to drop him off when he realized the firefighter wasn't responding, just staring straight ahead. "I turned around and there it was. It was basically coming straight at us," he said. "The main base of the tornado was over the woods. There

BEFORE: Boyer residence, 802 Highway 57, Parkersburg.
COURTESY BUTLER COUNTY ASSESSORS OFFICE

AFTER: Boyer residence, 802 Highway 57, Parkersburg.
COURTESY *DES MOINES REGISTER*

were two finger tornadoes dropping down. One of the finger tornadoes was going right for his [Ballhagen's] brother's house and his son's house. I know he was thinking about his grandsons. I was thinking I had to warn people and what I needed to do as soon as this was over."

In town, a weather spotter radioed a confirmed tornado was on its way. The warning siren sounded. But after only thirty seconds the siren fell quiet—a power failure had cut the electricity. Desperate to get word out, firefighters piled into trucks and raced up and down the streets, sounding sirens and warning people over the trucks' PA systems.

Meanwhile, Schipper and Ballhagen had cranked their truck around and were heading back toward town with sirens screeching. "I had a firefighter calling me and saying 'It's a minute before it hits town . . . thirty seconds before it hits town' . . . I said, 'I know, I know . . . I am right in the middle of it,'" said Schipper.

Schipper and Ballhagen pulled into the New Hartford fire station. Schipper expected to find his youngest daughter and ex-wife there; when he couldn't, he ran down the street looking for them. "I saw a big rush of dust like a huge cloud. Then a kid's jungle gym came flipping end over end down the middle of the street. Trees were snapping off. I thought it was the beginning of the end," he recalled.

The center of the tornado passed just north of New Hartford, sparing the downtown. Schipper returned to the fire station and found his daughter safe. He focused on recovery efforts. The twister had wreaked devastation on a residential area about a quarter mile north of town. Two people had been killed and approximately sixty-one homes destroyed.

The impacted area was heavily wooded. Downed trees and branches prevented rescuers from reaching the victims. "We had to round up chainsaws and find farmers with tractors. Then the trees had to be cut in small enough sections for the tractors to push them out of the way," Schipper said. "It took us forty minutes to get to what should have taken us three minutes. When you are dealing with an emergency, it seems like an eternity."

With reports of trapped people coming in, Schipper put out the call for help. But all emergency crews were being diverted to Parkersburg, and, in fact, his best ambulance was also helping those victims. "We were short on gear. We basically had a 1978 ambulance with a cot, and not

BEFORE: Scallon residence, 805 Hilltop, Parkersburg.
COURTESY BUTLER COUNTY ASSESSORS OFFICE

AFTER: Scallon residence, 805 Hilltop, Parkersburg.
COURTESY *DES MOINES REGISTER*

Sticks and stones—Parkersburg. COURTESY DAN BLEVINS

really a lot of stuff. We had to use a lawn chair to carry an elderly man up and over a hill. We set him down and I realized the sun was shining right in his face. George was his name. He had been trapped for forty minutes. I asked him if he wanted us to move him and he said, 'No, just let the sun shine on my face'."

Finally Schipper's calls for help got through, and rescue workers from Shell Rock, Janesville, and Hudson came to help. The work ground to a halt at midnight. Schipper went back to the fire station and organized a brigade of four-wheel-drive vehicles to search for people who possibly had been thrown into the woods.

He worked through the night and into the next day, not sleeping until Tuesday morning. There are long gaps of time he doesn't remember. Other memories are of trying to cope with the "never expected." Schipper has been on the volunteer fire department for eighteen years, eight of those as chief. "Nothing," he said, "can prepare you for anything like this. It is just a lot of reaction. I would have two cell phones and two people asking me what they needed to do. It got accomplished. The most difficult part of being the volunteer fire chief in a small community is that you know everyone and so these are your friends who are hurt and devastated. I am used to dealing with my friends one at a time. I am not used to fifty or sixty of my friends looking at me and asking me what to do."

ALL IN THE FAMILY

It's a family affair for the Truax clan to respond to emergencies in the Parkersburg and Aplington area. Rod Truax Sr. is an emergency medical technician and a Parkersburg firefighter. His wife, Beverly, is on the Aplington Ambulance Service and a dispatcher for Butler County. Truax's two sons, Rod Jr. and Rusty, are also Parkersburg firefighters. His third son, Reggie, is with the Aplington Ambulance Service. On that fateful day in Parkersburg, all five Truaxes played a role in providing comfort and care to the victims of the disaster.

Rod Jr. had been on an errand for the fire department when he was called in as a weather spotter. The afternoon skies had turned uncertain but not overly threatening when he checked in at the station. The atmosphere was relaxed. No one expected any trouble. The severe weather appeared to be heading north. In fact, Second Assistant Chief Ryan Siems told Truax and firefighter Mike Siems to take the grass truck, an old Dodge used for grass fires that lately had been plagued with mechanical troubles. The chief wanted to see if recent repairs were working. "Nothing is going to happen anyway," said Chief Siems, who admitted those words have come back to haunt him. Truax and Siems drove south of town and, sure enough, first the windshield wipers quit and then the engine died. They nursed the truck back to the fire station and awaited orders.

Minutes later, an Aplington firefighter radioed the Butler County Sheriff's Department. A tornado was spotted south of Aplington, headed right for Parkersburg. Siems gave the order to sound the town's new tornado siren. Inside the fire station, the six firefighters stood still, listening hard. The system had been installed and tested just two days earlier. The siren screamed out at exactly 4:54 P.M., and the firefighters erupted into action, grabbing gear and piling into trucks. They hurtled down the street with their own sirens blasting out a warning.

Truax took the wheel of his pumper truck; firefighter Eric Schoneman rode shotgun. They headed to the southwestern end of the town and spotted a man fueling his truck at the Kwik Star. "Get inside," Schoneman screamed as they roared by. Closing in on the town's lumber yard, they turned and nearly drove into the largest, ugliest tornado either of them had ever seen. The tornado was approximately 1,000 yards in front of them, furiously twisting and rotating as it churned through an open field. "It was throwing things left and right, taking up that entire field . . . it was about three-quarters of a mile wide," said Truax. A high-pitched hiss like a jet engine could be heard, but Truax was "so in awe of it, I wasn't paying attention." He slammed on the brakes and forced the truck to the north before making a snap decision to turn around and head back south to see where the monster was going. In doing so, he and Schoneman put themselves back in the path of the storm. Next to him, Schoneman was shouting, "Go faster!" but Truax already had the accelerator pressed to the floor. The tornado was moving fast, like a montrous black bird with its wings flapping. Schoneman hollered, "It's coming right at us!" but there was nothing he could do. The truck refused to go any faster. Truax fought to keep the vehicle from crashing. Schoneman opened his cell phone and began to videotape the tornado pounding across the road. "I kept asking him, 'Is it okay? Is it okay?'" recounted Truax. "He said, 'Not yet, not yet.'" The race against the tornado felt like hours, but it was only a few moments later when Schoneman finally said, "It's okay. We're safe." The tornado had crossed the road and was moving away from them.

The two men thought the worst was over. Truax called his wife and told her, "It's okay . . . it's gone." They headed back to town and came up over the top of a hill, and Truax turned to Schoneman and said, "Oh my God, it is not okay." He immediately radioed command and told them to "Get as many people here as you possibly can . . . our town is gone."

Coming into town, Truax and Schoneman were met by a team of storm chasers who had been following the tornado. "There was a large gas leak, and one of the chasers came up to our truck and was very assertive with me. I told him MidAmerican had been contacted. That's when everything just clicked. I just went right to work . . . right to rescue mode,"
said Truax. The pumper managed to squeeze under downed power lines and reach the edge of town. Immediately, citizens surrounded the rescue crew. A lady was trapped in the basement of her house. Truax ran to the house and called to the woman, but there was no response. Another man pulled up to him. "Come with me," he begged. "A woman is trapped underneath her car at the Kwik Star." All around Truax was the hiss and smell of leaking gas. "I headed back to the pumper for rescue equipment. Second Assistant Police Chief Ryan Siems was there," said Truax. "He told me rescue workers were on the way. By now I had lost track of Eric. But I can't explain it. All of a sudden these guys just started showing up to help. It was so unbelievable how help got there so fast."

Nowhere to go—Kwik Star in Parkersburg. COURTESY DAN BLEVINS

Another firefighter headed to the house to rescue the trapped woman. Truax drove the pumper truck over downed wires, chunks of homes, and torn-up trees, trying to reach the Kwik Star. A man, his head covered with blood and holding his left arm to his side, ran up to the truck. "Help him," Truax told the fire chief from Dumont. At the Kwik Star he found firefighter Brian Graham. The woman under the car was alert and talking. "I am fine," she said. "I just need to get this car off my leg." There was no time to lose, the woman was obviously badly injured but the rescue equipment was stuck in a vehicle half a mile away. Truax put his arms behind his back, grabbed hold of the back passenger wheel, and lifted the back end of the car. "It didn't even feel like I had lifted it. I didn't think I did but then I saw Brian pulling her out right on top of him. He was covered with blood. A couple of people ran up to me and asked if they could help. I said, 'Help me hold this car 'cause it's heavy,'" said Truax. While Graham got the woman stabilized, Truax next tried to free a man stuck inside his overturned car. But he needed the jaws of life and raced over to the nearest pumper. When he got back, the man was out. "A dozen or so citizens had flipped the car over with the guy in it. Fortunately, the man was okay," he said.

A moment later, a man who lived next to the Kwik Star collapsed after suffering a major heart attack. Truax rushed over to help perform CPR and found his cousin Kris Lind working to save the man's life. His father, Rod Sr. joined them, but the man could not be revived and was pronounced dead at the scene.

Next, Truax came upon an elderly woman sitting in her rocking chair with her destroyed home behind her. "Are you hurt?" asked Truax. The woman shook her head no, and Truax told her it would be safer for her to go downtown where there were people who could help her. "She looked up at me and said, 'I can't walk,' and I said, 'I'll carry you,'" remembered Truax.

The long night stretched on. "It gets kind of blurry now," said Truax. "We had been on the go for a long time." When the sun finally rose the next morning, Truax was still on duty. Stationed at one of the sectors in town, he helped residents search through their homes for any belongings. In the afternoon, he helped direct other rescue workers from outlying communities. He finally slept on Tuesday morning. He had been awake for two straight days. "We have asked ourselves a lot of 'what if' questions and it makes you sick to your stomach to think what could have gone wrong. But it didn't. I can't explain it. It all just worked out. I have never been so proud to be a firefighter, and a firefighter in the state of Iowa," he said. "I just get goosebumps when I think about it."

ANGEL ON EARTH

Susan and Paul Morgensen stood in the doorway of their home northwest of New Hartford and watched the EF5 tornado. It was half a mile away and appeared to be advancing right at them. "I have never felt such fear," said Susan. "I was scared to death." The Morgensen home is surrounded

For spacious skies—the Parkersburg horizon. COURTESY DAN BLEVINS

by trees, but still the immense twister was easily visible high above them. "It was way up in the sky as high as you could see," said Susan. "I said 'Oh my God, it is headed right to us.'" The couple raced to the basement for protection. Outside, the massive tornado flirted with the edge of their property but, aside from bringing down a few trees, shifted its route to a housing division just a short distance away.

Paul and Susan emerged from the basement and found to their relief that the twister's impact had been minimal. That relief was short-lived. They climbed on their four-wheeler to check on the path of the storm. They drove to the top of a small hill and were stunned by what lay before them. "It looked like bombs had been dropped," said Susan. "Everywhere was total destruction, with homes gone, wires down, and stripped trees."

They drove down to the hardest-hit area and met firefighter Ron Schipper, who asked if he could take the four-wheeler to check for victims down the road. Paul immediately agreed and left to go home for chainsaws to help clear the road. Susan, a registered nurse, jumped on the back of the ATV. Schipper was grateful to have her along.

Devastation drive—Highway 57 in Parkersburg. COURTESY JUSTIN STOCKDALE

> The biggest area of a recorded tornado was nearly two and a half miles wide. It occurred near Hallam, Nebraska, on May 22, 2004. It is important to realize that size does not necessarily imply strength. Large tornadoes can have meager wind speeds.

For all the world to see—Parkersburg. COURTESY DAN BLEVINS

When Susan and Schipper reached what was left of the first home, they found two victims. Norman Beuthine and his partner, Renee Kincaid, lay in a field of mud strewn with fallen trees stripped bare of leaves, bark, and branches. Beuthine had been killed in the storm. Schipper, who had worked for nearly twelve years alongside him, will never forget that horrible discovery. "It's hard to find a buddy like that and know he's gone. It was a long mile to search after you find your best buddy like that. It's still hard," said Schipper.

"Renee was lying flat on her back," recalled Susan. "It was really muddy and there was debris all around. The clothes had been stripped off of her so I covered her up with clothes and materials I could find. She was so covered with mud you couldn't even tell what color she was. Her face was badly swollen, she had bad cuts on her legs, and one of her legs was lying in a weird position. I asked her to squeeze my hand and wiggle her toes. She could do that so I knew she wasn't paralyzed. She kept saying, 'Don't leave me. Don't leave me.' She asked about Norman and I said, 'We are looking out for him.' Norman was lying about twenty feet from her. There were two dead horses right beyond him. I felt so useless because I had nothing to help her with . . . no bandages or anything," said Susan.

While Schipper continued down the road to search for more victims, Susan stayed with Renee, comforting her. She could hear chainsaws in the distance, so she knew help was on its way. Still, it felt like an eternity before rescue workers arrived with an ambulance. Paramedics put a neck collar on her and gave her oxygen before carefully lifting Renee onto a backboard, but because the field was so muddy, she had to be carefully transported to the ambulance on the back of a John Deere Gator. Susan was asked to accompany Renee in the ambulance to the hospital. "We drove down 325th and I just couldn't believe it. Every house was wiped out. It was like something you see in a movie. The ambulance was weaving around to avoid the fallen trees and people chainsawing. Two horses were just walking down the road, dazed. I kept Renee talking to us the whole way to the hospital," said Susan.

After seeing Renee settled in the intensive care unit in Cedar Falls, Susan and the paramedics returned to New Hartford. It was pouring rain, and rescue workers were slogging through mud and debris. Susan joined in alongside Paul, but after a while it became apparent there wasn't much more to do until the rain let up. They returned home and took hot showers. "We were soaking wet and covered with mud and blood. I felt so guilty being able to take a shower when other people didn't even have a home left," she said.

Even with her training as a nurse, Susan said, nothing can ever prepare you for what happened. "It really rocked our world and changed us. We look at things a little differently. We have talked to many people who lost everything, and they say, 'Yes, but we are alive.' Paul and I often think about how lucky we are," she said.

Two days later, Susan returned to the hospital. She had been thinking so much about Renee and hoped her condition was improving. "I held her hand and said, 'This is Susan. Do you remember me? I was with you after the storm.' She said, 'Oh yeah,' and squeezed my hand. She was in intensive care so I didn't want to stay too long. I walked out of her room and her sister gave me a huge hug and said, 'You are her angel on earth.'"

FIRST TIME FOR EVERYTHING

Brian Graham wanted a great place to raise a family, so he moved his family to Parkersburg about ten years ago. It was the ideal setting, with a good school district, family nearby, and a more economical lifestyle.

When he was asked to join the volunteer fire department, Graham's first reaction was, "Thanks, but no." It wasn't until he injured his knee and had to decline a job with the Iowa State Patrol that Graham changed his mind.

Graham passed his firefighter test in March 2008. His first call was to a hay barn on fire, his second an EF5 tornado. "The day kind of started like most Sundays do. Did the family thing together, went to church," he said. Later in the afternoon, Graham noticed the sky was looking a little odd. He visited with a neighbor and commented on how the firefighters hadn't been called out much for weather spotting. Shortly after, his pager went off. He was needed immediately. "Thought we would go out and watch the rain as usual," he said. Graham teamed up with another firefighter and headed west. But the fire station radioed and said the storm was going north, so they changed direction. "We were sitting up high on top of a hill. The wind was really blowing and it started to rain. Then we started getting hail about the size of a dime or nickel. Then we heard someone had spotted a tornado and it was coming in from Aplington. Then the radio blew up with traffic and everyone talking."

The two headed back into town. As they drew closer, a funnel cloud popped up right on top of them but moved off quickly. "I couldn't figure out why everyone sounded so hysterical on the radio, because the north side of town looked just fine. Then we got midway through town and came upon the area where the tornado had gone through. It was just

Room with a view—looking east through Parkersburg. COURTESY DAN BLEVINS

devastating what we saw. It had leveled everything. Then we came up around a curve heading toward the Kwik Star. There was a cow in the middle of the road, alive but critically hurt. Several people ran up to our truck and said a lady was trapped under a car," he said.

The only way to get through the damage was on foot. So Graham got out of the truck and left his colleague to find rescue tools. "I came up on the lady who was underneath a car. She was responsive. So the first thing I could think of was to hold her head and talk to her and make sure she was still with us," said Graham.

He realized she needed to be moved right away. Summoning help from another woman, Graham asked her to stay with the victim while he hurried to find another firefighter. He said, "I probably ran a quarter of a mile and found Rod Truax, who was in a fire truck. We were able to drive over the debris and downed wire to get back to the Kwik Star. There was a heavy, heavy smell of natural gas. That was right across from the Kwik Star. You could smell fuel and natural gas and it was pretty disturbing. There was a guy stuck in a car by the gas pumps and the car was on its side. Rod and I checked on him. He seemed pretty secure so we got back to the lady under the car. She was in a more critical state. Rod said, 'I think it's time to get her out.' Rod grabbed right above the rear wheel and pushed the car high enough for me to pull the lady out and make sure she was safe. Some citizens came over and told us the guy in the car had gotten free under his own power and they pushed the car back on its wheels."

After making sure someone could stay with the victims, Graham headed down the block to look for people trapped in their homes. "I know we walked around town on the south side for a long time, but it seemed like it was in a blink of an eye," he said.

During the search, a man came up to Graham and asked if he could go down in his basement to rescue his little Scotch terrier. "His home was shifted off the foundation. So I went into the kitchen and we climbed over the top of the debris. The dog was safe. The man had lost everything else, but now his wife, children, and his dog were safe. I think that was a big relief for him," he said.

Graham and the rest of the firefighters remained on duty throughout the night. "All of the guys I have talked to have said the same thing: 'We lost time. Little areas are only blurs. I don't think I am an extraordinary person, but when put in extraordinary circumstances you do what you have to do to help your own town.'"

JAW DROPPING

Eve Halligan studied meteorology and climatology at the University of Nebraska. As a graduate student, she had chased storms in Kansas and Oklahoma. While she saw plenty of wall clouds with ominous rotation, she had never seen a tornado.

The extraordinary becomes ordinary—a pummeled Chevy van.
COURTESY JUSTIN STOCKDALE

On the morning of May 25, 2008, she called her sister, who lives between Aplington and Parkersburg, to tell her to keep an eye on the sky. "I knew we were going to have severe weather that day and my sister doesn't have a basement," Halligan said.

Halligan and her two young children were en route to pick up her parents about eight miles northwest of New Hartford. From there, the family headed into Waterloo to observe Memorial Day and visit the grave sites of relatives.

On their way home, dark clouds began stacking up on the western horizon. "I turned on the radio," remembered Halligan. "There were tornado warnings but the threat seemed to be more to the northeast of where we were."

Worried again about her sister and family, Halligan called to warn them about the weather. Instead, her sister told her the storm was headed her way. "My sister and her family actually watched the tornado come out of the clouds and touch down on the ground within a mile of their house," said Halligan. "They got into their vehicle and took off," she added.

North of New Hartford, Halligan and her parents drove up a large hill and came face to face with the killer storm. "It was one of those moments that make your jaw drop a little," she said. "I never in my life thought I would see a tornado like that." The tornado was approximately two and a half miles away, but advancing quickly. "I was thinking 'oh crap' because I knew what it was and that it was coming fast toward us," she said. A trained meteorologist, Halligan realized they were seeing one of the largest tornadoes imaginable. She turned to her parents and said, "I think at the very least it is an EF4." Inside the car, Halligan's nearly six-year-old son hugged her arm and said, "Mommy, I'm scared. I'm scared." While the north end of the storm was rain-wrapped, Halligan and her family could clearly see the south edge. "There was a smaller funnel that kept dropping down and touching briefly and then going back up," Halligan said. "When I saw that, I yelled to my Dad to stop, turn around, and head to their best friend's house for shelter." Driving in a heavy downpour with golf ball–sized hail pounding down, Halligan and her family cleared the cemetery hill area in New Hartford with only five minutes to spare. No one was at their friend's home, so they decided to outrun the storm.

A world turned upside down—the remains of Parkersburg.
COURTESY JUSTIN STOCKDALE

They swung onto a gravel road and drove away. "If it had gotten closer, we would have stopped and sought shelter," Halligan said.

Once out of the path of danger, the family headed to Halligan's house south of Parkersburg. She left the children safe at home and drove back toward Parkersburg, staying about a mile out of town so as not to interfere with emergency efforts. "But even from a mile away we could see the devastation, and it was pretty shocking."

Outrunning an EF5 tornado has changed Halligan's outlook on storm chasing. "I have to say it has taken away my desire to chase storms," she said. "We live near Parkersburg. I used to go to that grocery store and the Kwik Star . . . it's too painful to see what they do."

IN HIS OWN WORDS

Miles Humphrey flew in two wars as a fighter pilot for the U.S. Marine Corps. He is the father of nine children, now all grown and married. He is the kind of individual you would want nearby in an emergency because he is simply unflappable. On the day of that EF5 tornado, Humphrey stood outside his home on the outskirts of town and calmly watched the tornado develop from a "tiny gray smudge" in the sky to a full-blown killing machine. "At first, it didn't go right or left, that smudge just kept getting bigger. Then it connected the bottom of the cloud with the horizon. But it was sunny all around it. I finally woke up to the fact it was a tornado," said Humphrey.

The faint sound of a siren confirmed the tornado, and Humphrey told his wife it was time to leave. "She said, 'Let's get in the basement' and I said, 'No, I can see it's coming right at us. We have to leave,'" said Humphrey. They drove east, intending to make it to Highway 14 and turn south. "As we hurried to get there, I could see the tornado was entering the west side of Parkersburg, so I hurried and drove seventy to seventy-five miles an hour. We made it to Sinclair Road, which is about two miles east of Parkersburg, and turned south for about three blocks and waited. Debris started flying around the car. It was like a snowstorm. Then I could see the tornado plow through Parkersburg. It was a large black mass like a pillar. It looked like it had a funnel swimming around inside, and it was proceeding on a steady easterly course. It was actually moving rather leisurely, just slowly grinding along. I wasn't frightened . . . actually I was taking pictures. We could see quite a bit of debris falling and moving when the tornado hit the Sinclair grain elevator. The grain bins were torn. We watched it go through there and then we drove back to Highway 57 and I was going to go to the Sinclair elevator but two big trees fell down and I had to quit. We did see white puffs of ammonia coming out of the anhydrous tanks."

By dawn's early light—picking up the pieces of Parkersburg. COURTESY SCOTT TJABRING

Humphrey and his wife hurried to their son's home in town. At one point, they had to abandon their car and walk because of the debris littering the street. His son was standing in front of what used to be his home. Finding everyone safe, Humphrey continued through the devastation to check on his friends. "I was headed toward the Luhrings and discovered later I had walked right past the body of Richard Mulder. Before I reached Luhring's house, I came upon the body of Mrs. Mulder. She didn't have much clothing on her and was obviously no longer alive. I walked up to Herman Luhring and he was all covered with blood. There was a young man standing on each side of him, and they looked like they were ready to catch him if he should fall. The moment he saw me he broke into a big smile and said, 'Miles I am so glad to see you.' I knew then that he was probably going to make it. I asked about his wife, Shirley, and one of the young men said they hadn't found her yet."

"So I left then," Humphrey continued, "and turned around and one of the Mulder boys had found his mother. It was a sad scene and I didn't want to intrude so I went down the street. On Fourth Street the scenes were incredible. I came across a motorcycle that looked like someone had taken a giant hand and just squeezed it. Mountains of debris were on each side of the street. There were no houses and no trees. I met another couple; they had totally lost their house, but they did not go through the tornado because they had driven in right after it. They were astounded and bewildered by what they saw. I kept going farther and saw another family who had lost their house, but the garage was intact. I asked them if they were okay and they said yes. So then I came to our parsonage. The pastor had just managed to come up from the basement. Her left arm was in a sling and she looked as if she was in shock. I walked up to her and we hugged. Then she asked, 'Do you see a wedding dress?' I said no, but there were all kinds of blankets up in the trees." The dress had been purchased in France and was for her daughter's wedding in a few weeks. "We went and looked in the closet in the heart of the kitchen—two sides were still standing. Everything was in it except for the wedding dress.

"Then the pastor and I walked slowly up the street to our Methodist church. It was standing, but most of the shingles were gone and all the windows in the fellowship hall were blown out.

"My wife and I continued to look at things and for people until sunset. By this time there were a lot of people around, and the rescue workers wanted to instill a curfew at eight. So we went home. Our house was standing but there were trees down, windows broken, and shingles missing. Our eighty acres of farmland were strewn with debris.

"I have been taught to handle tough situations so my advice to anyone who wants to try and outrun a tornado is, don't do it. If you don't know the country, you could get caught and you won't be able to escape."

THE LAST SUNDAY

Bethel Lutheran, the community's largest church, is a landmark in Parkersburg, Iowa. Its stone walls have stood the test of time. That ill-fated day was Pastor Donna Ruggle's last Sunday. As the interim minister, she had helped fill a gap in the church's ministry until a permanent replacement could be found. It had been a difficult eight months, with an unusual number of funerals—including one for an infant. But a new minister would be arriving in a few short months, and Ruggles carefully chose her scripture and hymn for what she thought was her last service. She finally selected Psalm 73:23: "Yet I am always with you; you hold me by my right hand," and the hymn, "Never Fear Little Flock."

After the service and a thank-you reception, Ruggles drove the forty miles home to Waverly. Later that afternoon, she called a pastor in Charles City and was greeted with the question, "Why aren't you in Parkersburg?" The pastor relayed the information about a massive tornado striking the town, and Ruggles quickly changed back into her clerical shirt and retraced her path to Parkersburg. At the city limits, Ruggles, a native of Kansas, "knew immediately it had been an EF4 or 5" because of the scope of the damage. She joined a long line of cars waiting to enter Parkersburg. A police officer was stopping every vehicle, and, when she finally made it to the head of the line, he told her, "I doubt you have a church left."

Ruggles had arrived less than two hours after the tornado hit, and everywhere she looked she saw "people just walking around like zombies, just dazed. My first thought was 'Oh God you put me here because

I know how to do this,'" said Ruggles. "I kept saying to myself, 'you need to deescalate the crisis.'" With a background in hospital chaplaincy, Ruggles had worked with many death and dying situations and knew how to handle a crisis. She opened the church for victims and staffed telephones before heading over to the American Veterans' Center to help wrap the elderly residents in blankets and load them onto buses to get them to shelters.

Ruggles' church was not structurally affected. The roof was damaged, there was glass everywhere, and the cross was slumped on the church's steeple. Inside a recessed area on the church's exterior, Ruggles found a dazed owl slowly blinking its eyes.

Crossroads of the devil—the resurrection of Parkersburg.
COURTESY PASTOR ARNOLD FLATER

The congregation was much harder hit, with forty-four homes destroyed. Ruggles set up the church as a community center, offering information and assistance. "The helping agencies were surprised by how much we did ourselves," she said. "But these people are Iowans and they don't want charity. They will take care of themselves."

STRIKE TWO

Rich and Patti Schmitz live on a farm approximately twenty miles east of New Hartford, Iowa. When the couple bought the estate fourteen years ago, they tore down the hundred-year-old farmhouse, built a new home, and moved in with their growing family. In 2000 the unexpected happened. While attending a school function for their eldest child, they heard tornado sirens screaming in the distance. Patti called home and told the three younger children, ages twelve, ten, and six, to head to the basement. The three obeyed and, minutes later, what is called the Dunkerton tornado slammed into the new farmhouse, destroying it. The children, hiding beneath a pool table in the basement, were safe. Undaunted, the Schmitzes rebuilt the house on the intact foundation, using the exact same plans.

On Sunday, May 25, 2008, the family was sitting down to dinner when news reached them of a huge tornado tearing into Parkersburg. "My husband said, 'Let's load up the chainsaw and see if we can help,'" said Patti. "We knew what it was like to go through something like that." Ten minutes later, plans abruptly changed. A tornado was spotted just a few miles from the Schmitz's home; Patti, Rich, and the children rushed to the basement. The kids took shelter under the same pool table that had protected them eight years ago. "I could see out an egress window," said Patti. "It wasn't like a tornado, it was like a huge wall of dirt coming at you. You could see the dirt and corncobs swirling around. That's when Rich and I got under the pool table, too."

The Schmitz's home was once again targeted by a tornado. "We looked at each other and said, 'No, this couldn't be happening again,'" said Patti. She described the tornado's fury as a wild party underway with chairs being dragged all over the floor. While the destruction took only minutes, Patti said it "felt like an eternity." The horrible noise finally subsided, and Patti and Rich looked out the egress window again. This time debris from the garage was scattered everywhere. "We knew that couldn't be good if the debris from the garage was sitting in our window," said Patti. Emerging from the basement, they found the house was again destroyed, and what little remained was filthy. Every square foot was strewn with corncobs, dirt, and insulation. "The first time we built," said Patti, "we used fiberglass insulation, but the second time we used an insulation that was like ground-up paper, and it was everywhere. All the drawers and cupboards, even the washing machine and drier, had been popped open, filled with ground-up paper, and closed again." A huge machine shed used to store a backhoe, tractor, and grain carriers had disappeared. The backhoe was discovered in a knoll behind the house. The one exception to the incredible damage was the vineyard. The couple has 140 vines they tend and, while the leaves looked a little sickly, the vines had stayed in the ground. "There are some new leaves coming back," said Patti. "So they may be okay, just delayed a year."

The Schmitzes are now working on plans to build a third house on the exact same foundation. This time, however, the home will be single story, and there is talk of looking into Hurricane Katrina–type plans that have supports in place to withstand winds up to 200 miles per hour. The advice from friends has included building everything from an earth home to putting in additional concrete walls.

The irony is that the original farmhouse had stood for a hundred years without any major damage. "It is a little scary to rebuild," admitted Patti. "I hope we are doing the right thing."

MILK, BREAD, AND A TORNADO?

Linda Blevins, her husband, and elderly father from Arkansas arrived in Parkersburg late Sunday afternoon, approximately twelve minutes ahead of an EF5 tornado. The trio was returning from the airport in Des Moines after picking up Linda's father. Linda's son was getting married, and she had a full plate of things to do. "On the way home, it looked a little strange but I thought, don't worry it's just a rain cloud," said Blevins. Arriving at their century-old home, Blevins' main concern was getting her father settled. She took his suitcase upstairs, and when she came back down, the television was reporting a tornado warning. "I saw there was about five minutes left of a tornado warning and thought, I am sixty-one years old, that will never happen," she said. Blevins jumped into her car and headed to the grocery store to pick up dinner. A nervous grocery clerk greeted her and repeated the tornado warning. "She was just shaking," recalled Blevins. "She said she wanted to lock the door but I said, 'No, no, I need my groceries.'" Blevins grabbed a cart and started shopping when the store lights flickered off. She hustled over to the checkout lane, where the frightened clerk told her the conveyer belt wasn't working. "Don't worry," Blevins told her, "I'll just push the groceries through." While Blevins wrote a check for the groceries, another store employee yelled that a tornado was coming. "Take shelter in the back of the store," he urged. Blevins and a small group of store employees and customers climbed into a store cooler and slammed the door just as the tornado crashed in. "I barely counted to ten and it hit," said Blevins. She remained calm and recalled that in thirty seconds it was all over with.

No one was prepared for what they saw when they finally pried open the cooler door and squeezed through the opening. "One guy said, 'There

Standing strong—battered homes braved the storm. COURTESY HEATHER KRUGER

is no Parkersburg left anymore,'" recalled Blevins. The grocery store was destroyed. A few products could be salvaged. Ironically, Blevins' hamburger and lasagna still sat on the checkout counter.

The thing that struck her most, said Blevins, was the silence. Nothing moved for a few moments. Then she heard sirens wailing. Worried, Blevins raced home to check on her husband and father. "My house was just a shell," she said. Miraculously, her father had escaped harm by taking shelter in a small bathroom. Her husband, unable to open the bathroom door slammed shut by the storm, had lain down by the dishwasher. Blevins' next concern was her next-door neighbor, seventy-four-year-old Ray Meyocks. "I think he was in the basement, down in the rubble," she said. "I could see him looking up at the sky. He knew my voice and he knew people were there to help him." Meyocks had a pacemaker for his heart, and, despite the heroic efforts of paramedics, he collapsed and died shortly after being rescued. "His grandson Carl played with my boys," said Blevins. "It is unbelievably sad."

The roof was gone from Blevins' two-story home, and walls were partially sheared off. "I could stand in the middle of my tub and see two miles of devastation," she remembered. A few scratched-up photographs and bits of clothing were salvaged. A few days later, Blevins' son was married, and she wore the dress she had purchased for the wedding. It was in a closet, perfectly preserved, still in its plastic sleeve.

HANGING ON FOR DEAR LIFE . . . LITERALLY

Tom and Sue Teeple work from what could be called the unofficial nerve center of Parkersburg: the barbershop. "I am the town's beautician and he's the barber," said Sue. The couple planted roots in town in 1977 and hasn't budged since. "I am a Parkersburg man," said Tom with pride. On the day of the EF5 tornado, Tom and Sue had settled into the living room to watch the races on television. When storm warnings interrupted the programming, they were impatient for them to be over. But then things changed. "It started to look pretty icky outside," recalled Sue. So Tom went out and watched the sky with his neighbor. After a moment or two Tom saw the tornado. "It was ugly. It was a big black cloud with a wide, wide tail," he said. Tom raced back inside, thinking only of his wife's safety. Sue has multiple sclerosis and can get about only with the aid of a wheelchair, walker, or scooter. Tom was barely able to get Sue settled on the landing that leads in one direction to the garage and the other to the basement. "Jo Dreesman, a next-door neighbor, was on her knees beside Sue, holding her," Tom said. He quickly threw a quilt over the two women and grabbed the safety rails installed for Sue. No sooner had he bent over to try to protect the women than the twister exploded into the house. "It was the most terrible sound in the world," said Tom. "It was like a freight train and it kept getting louder and louder," he added. For ninety seconds the tornado whipped around him while Tom held on to the safety rails, fighting for his life. The twister pulled so hard, trying to suck his body out, that his legs were lifted and dragged straight back behind him. "I lifted my head and saw the south end of my house going

Mowed down—a lawn mower adorns a Parkersburg tree.
COURTESY HEATHER KRUGER

up in the air. The second time I looked up the kitchen was going," he said. Just when Tom's strength was nearly gone and he felt himself giving in to the power of the storm, it ended. "It was like a light switch had been flipped. It was just over," he said.

Tom's body had shielded both women so well, neither had been harmed during the twister. Miraculously, Tom had not been injured either. "When that storm hit, I went into a shell. I am telling you the force of that thing was incredible. I am built like a rock, I am pretty strong, and that thing could do anything it wanted. Of the eight deaths, four of them were people sucked out of their homes," he said.

The next thing Tom knew, five or six people had come through the wreckage calling to see if he and Sue were all right. Sue was carried out of the rubble and Jo went to check on her family. Tom stood in the middle of the devastation and remembers an awful smell. "It was the smell of natural gas, and it was a terrible smell," he said. "I am truly humble and thankful that we are alive."

Tom is the second in his family to live through a tornado. In 1968 an EF5 tornado swept through Charles City, Iowa, leaving his older brother with nothing. "He lost everything," remembered Tom. "I never thought I would live through it. There is an old wives' tale the old-timers in town tell, that a tornado will never hit Parkersburg because of the fork in Beaver Creek. Well, I got news for you, old-timers, the fork changed."

IT'S JUST SO GOOD TO BE ALIVE

At quarter to five on that Sunday afternoon, Marty Luhring climbed into the shower to clean up before a graduation party. Her eighteen-year-old son John was about to head down the road to Aplington, approximately eight miles west, to attend another party. Outside, her husband, Larry, was watching dark, ominous clouds gathering just to the south of Aplington. The day had been beautiful, but a menacing wind was beginning to blow. John left in his truck and Larry continued to watch the storm. Within minutes the storm had amassed an awesome strength, growing more than a mile wide. Larry realized he and his wife were standing directly in the path of the biggest, ugliest tornado he had ever seen. He quickly picked up his cell phone and called John to warn him.

"John told me, 'You and Mom had better take cover, it is coming right for you,'" said Larry. He ran inside and called for his wife to hurry out of the shower. Marty grabbed her clothes and they headed for the basement. The house had lost power, so, feeling their way carefully in the dark, they hid behind a water heater and hugged each other tightly. Then there was nothing but a queasy silence. "I said to Larry, this is the quiet before the storm," said Marty. "You could have heard a pin drop—and then it took off like a lion and roared and roared." Debris began to pelt the windows. Larry hugged Marty tighter. "I said, we're going to have some trouble . . . hang on honey, we're in for a ride," remembered Larry. "And then there was one big 'Vroom.'" Marty shut her eyes as the tornado boomed overhead, tearing the home off its foundation and tossing it away like a dollhouse. Larry looked up and saw "a black, flat cloud, like a table"

The sounds of silence—Aplington/Parkersburg High School.
COURTESY JUSTIN STOCKDALE

moving overhead. "I never heard the train," said Marty. "But the pressure was so intense. My ears were plugged," she added. The intense suction of the tornado began pulling everything out of the basement, first the water heater and then Marty. "We were holding on to each other and the storm started pulling my left hip away and then my leg and I felt myself start to go. I said to Larry, 'The storm is starting to take me away from you,' and he said, 'I won't let you go.'" Larry gripped Marty more tightly and they both prayed. "When it started to pull me away, I said 'Hail Mary full of grace' and then the next moment we were pushed down into a sitting position. I know she was with us," said Marty.

Exposed in the gaping hole of what had been the basement, Marty and Larry were pelted by bricks, hay bales, and other debris. A tree fell into the hole just feet from where they were. All familiar landmarks had vanished, along with their farm. As the light gradually returned, they found the stairs. Together they climbed, holding hands. Marty found a golf shoe perched on a step and handed it to Larry, who had lost his left

Wipe out—the southern third of Parkersburg ceased to exist. COURTESY GAYLON ISLEY AND DENNY MILLS

shoe in the storm. It was a left shoe. "I said, 'You are not going to make *Gentlemen's Quarterly* but at least you'll have two shoes on,'" said Marty.

The devastation was heartbreaking. The house was gone. The new office building, pig shed, and barn were gone. The trees were stripped like toothpicks. The bodies of cows were stacked against an overturned van. Larry's head was bleeding from a deep gash. "It was a picture you never want to be part of," said Marty. "Larry said, 'My god we have lost everything,' and I said, 'No, we have each other and we are alive,'" said Marty. "It is so good to be alive."

I HAD A PREMONITION

In Parkersburg, the minister of the United Methodist Church is simply known as Pastor Betsy. Betsy Piette has served the Parkersburg congregation for five years. When parishioners talk about their minister, you can hear the smiles in their voices. For Pastor Betsy, that fateful Sunday afternoon began as usual with a morning service. Normally she would have stayed after the service and caught up on office work, but because she was recuperating from surgery she headed home to the parsonage.

Later that day, the pastor sensed a change in the peaceful afternoon. "I had a premonition that something was coming. It was just a sense that things were not normal," she said. Pastor Betsy turned on the television, and soon a tornado warning was announced. "Like most people, I don't pay much attention at first. But then I did something really odd. I thought the power might go out, so I heated a frozen dinner in the microwave and then found a flashlight. At a quarter to five I really began to feel uneasy, so I took my daughter's wedding dress and the flower girl dresses and put them in a hall closet," she said. As soon as she closed the closet door, the lights in her home began to flash. "I could hear the tornado sirens," said the pastor. She threw her purse in the pantry, grabbed a sleeping bag, and headed to the basement. "All of a sudden, it was quiet. Then there were the sounds of a train. That was a surprise, it really does sound like a freight train. Then you hear things start to snap and pop. My ears popped from the change in the pressure. I said a prayer, and it was all over in a minute. God was right there holding me. Live or die, it was going to be okay."

When Pastor Betsy came up from the basement, the stairs were covered with the entire contents of her refrigerator. Her kitchen still stood, but the corner of the house where the bedroom had been was stripped clean. There was nothing but dead silence all around her. Then the quiet was broken by the ringing of her cell phone—one of her parishioners calling to see if she was alright. Amazed that her phone still worked, Pastor Betsy left messages with her daughter and father to tell them she was safe. Emotionally battered, she broke down while leaving a message for her father. "My dad told me later it was the most gut wrenching message a father could hear. I kept saying, 'I don't know what to do. I don't know what to do,'" she said.

With people all around calling one another, the pastor lost track of time while reaching out to help. At one point, she walked up to the church to see if the building had been damaged. On the top of the church are three crosses. Looking up, she saw one of the three crosses badly bent, but all of them there. "I never felt so calm," she said.

POPPING BEAN JARS

Ever since she was a young girl, Shirley Siems has canned vegetables, first with her mother and later as a young wife and mother. She has put up countless jars of green beans, peas, carrots, and applesauce.

Shirley's base of operations in her Parkersburg home was a fully equipped kitchen in her basement, with shelf after shelf of neatly labeled glass jars filled with Shirley's work.

On May 25, 2008, Shirley and her husband, Raymond, were watching television in the upstairs living room. Approximately an hour before the storm hit, Shirley turned to Ray and said, "Do you hear those jars popping downstairs?" Ray replied he did but had no idea why the lids would be popping. The moment passed and the couple brushed off the unusual occurrence.

Minutes ticked by, and the weather outside grew dark and cloudy. The wind picked up. Suddenly, tornado sirens blasted through the air. The Siemses quickly picked up their dog Cody and headed to the basement where, tucked underneath the garage in a corner, was the cold-storage cellar Shirley used to store her canned vegetables. As they huddled in the cellar, the lids on the jars started popping again. This time dozens of lids popped from the building pressure of the rapidly approaching tornado. "Then it sounded like a train and then several trains," said Shirley. "A huge 'Whomf' followed, and then it was over." The couple slowly emerged from the tiny cellar and could see a little light above them. "We figured our garage was gone," said Shirley. But after climbing out of the rubble, they saw the house they had built and lived in for forty-three years was also gone. "It just took our breath away. We couldn't believe what had happened in a matter of seconds," she said.

In the clean-up that followed, Shirley's jars of vegetables were hauled upstairs from the cellar. "The lids had popped, but the seals were still good," she explained. But not for long. Once the jars hit the light of day, the vegetables quickly turned moldy from the bottom up. Even Shirley's dried fruit turned orange and sickly after being exposed to daylight. "It had turned into bacteria from the pressure of the tornado. I had never seen anything like that. We had to throw everything away," she said.

The Siemses bought a condominium in another town and moved. Still, the nightmare that destroyed their life in Parkersburg haunts them. "It is really scary," she said. "Even now, when the wind blows, you still get so frightened."

Easy come, easy go—the First State Bank building, Parkersburg.
COURTESY SCOTT TJABRING

A THOUSAND BLACK CLOUDS

As director of development for the Salvation Army, Sue Hennings knows how important financial support is to the health and well-being of a community. After years of providing hope and help to those who have lost their homes, she never expected to find herself in that same position. "It can happen to anyone," she said in a whisper of a voice. Hennings' lungs were damaged by the debris and carcinogens hanging in the air after Iowa's EF5 tornado crashed through Parkersburg.

She and her husband had spent that deadly afternoon planting summer annuals, a hot, sticky job in the warm weather. "Thunderstorm weather" she had commented to her husband. He had replied, "No, it's more like tornado weather," and she had called back, "Don't say that!"

At 4:00 P.M., Sue went inside to wash her hands and decide what to do about dinner. "I could look right out my back window and see the Aplington area," she said. "I didn't really see much." Then the NOAA radio alarm sounded the warning that a tornado had been spotted at the Butler-Grundy county line. Sue looked out the window again and saw birds in the back yard. She went back to work. "When I turned back for a second time, I saw this huge swirling mass of a thousand black clouds. It was enormous and fascinating to watch as other clouds joined and swirled in together. And it was moving very, very quickly," she said. As Sue grabbed her cell phone, she heard her husband yell for her to get to the basement. Moments later, tornado sirens started screaming. Fire and police sirens added their shrill alarms, and all that former quiet was brimming with fear and panic.

"We were downstairs for maybe two minutes when it hit," said Sue. "We held each other and prayed. The sound is horrendous. It is like a zillion gatling guns shooting at your house . . . nails spitting all around you. Then you could hear rafter after rafter cracking and windows breaking with tremendous booms and bangs. Dishes were breaking and two by fours battered the foundation. I thought the house was going to cave in on us."

After it passed, the two crawled out from the wreckage to a nightmare. "There were dead cows in the street, and kitties and dogs," Sue said. "Debris was piled up everywhere and everyone was looking for their family and friends."

While the Henningses managed to save a few personal mementos, their past had been obliterated in just seconds. "You have no idea of what you need because you have nothing," said Sue. "Our bank blew away, our credit cards are gone, the grocery store is gone. We have no cars. You are really helpless and there is nothing you can do."

The Henningses sought shelter with their son in Waverly, Iowa, but were quickly evacuated when that town flooded. As they now work to rebuild their lives, Sue sees firsthand how dollars are spent to help those in need. "I have worked for thirty years as a fund-raiser," she said, "and now I can really see how the money is being used."

CHANCE

Chance is a three-year-old yellow Labrador who, it turns out, was aptly named. Chance lived in Parkersburg with his family, Pam and Dan Bleeker and their five-year-old grandson Andrew.

On the day of the EF5 tornado, Pam Bleeker was barefoot, standing outside, when the wind started to pick up and the temperatures dropped. "I went inside and started looking out the window," said Pam. "I looked toward the west and a split second later, there it was. A huge black wall coming fast." Pam quickly grabbed Andrew and called to her husband, "We gotta get in the basement." Dan began calling for Chance and the other family dog, but the canines wouldn't budge.

The house started breaking apart as Dan raced down the basement steps. In the basement, Pam pushed Andrew beneath a canning table and held him tight. She had no idea if her husband had made it to safety. "It was extremely loud . . . an incredible roar," she said. Ten feet from Pam and her grandson stood the furnace and water heater. In seconds both were sucked out of the house, and Pam fought to keep Andrew safe. "Then the noise just stopped. It was all over with." She looked up and saw daylight through a hole. She pulled Andrew up to safety and began to call for her husband. "I just kept hollering for him. He was probably two feet from where we had been . . . pinned under some boards. I had no idea he had been that close to us," she said.

Outside, the family could see how a fallen wall had protected them from being crushed to death. Their van had landed upside down on top of the wall, which kept it from falling into the basement. Pam suffered a severe gash on her bare foot. "The doctor said it looked like a hatchet had split it," she said. Andrew was "walking in circles, his eyes as big as saucers." Pam also remembers a loud echo that kept booming in her ears when she spoke.

Dan and Pam Bleeker suffered another terrible loss in the tornado. Dan's mother, seventy-one-year-old Leasa Bleeker, lived across from the cemetery in New Hartford. "She went to bed really early and must have been sleeping when the storm hit," said Pam. The body of Mrs. Bleeker was discovered in her yard in a ditch. "She was found in her nightgown," said Pam. "Her home was blown away," and the contents of her home were in a "thousand pieces." Mrs. Bleeker's dog Penny survived the storm.

More heartbreak came when they realized their two dogs were gone.

Just another day in Parkersburg—sunset on the rubble. COURTESY SARA STEARNS

"Especially Chance, because he was a member of our family. We loved him so much," said Pam.

Nine days passed, and then a miracle came. Pam received a phone call from her insurance company. Agents were in town at the makeshift lot where damaged cars had been towed. "They asked me if I had a yellow dog, and I screamed 'Yes! Yes! Yes!'" remembered Pam, adding, "I was jumping for joy and shaking and crying at the same time." She called her husband, who had been working in town shutting down gas lines, and told him to hurry to the car lot.

Minutes later, Dan called with the crushing news. He had found his wrecked truck, but there was no dog inside. Pam was devastated.

At the car lot, Dan decided to check again and saw a group of people standing next to a car. "Has anyone seen a yellow dog?" he asked. At that moment, Dan saw Chance sitting in Pam's damaged car, refusing to respond to anyone. "He sat with his back to the steering wheel and wouldn't look right or left," said Pam. "My husband yelled 'Chance' and he nonchalantly turned his head. My husband started talking to him and then he bounded out of the vehicle and was everyone's friend," she added.

Chance had a couple of small cuts on his face, his ears were full of wood ticks, and he had lost twelve pounds. "Which he needed to do," said Pam with a smile. "He's actually at a healthier weight now."

The Bleekers say Chance is very happy to be home and always wants to be right next to them. Apparently this yellow lab isn't taking more chances.

CAN I COME IN?

In the cool of their basement family room, Virgil Goodrich and his wife, Dianne, had their television set tuned to KWWL-TV, the NBC affiliate station in Waterloo, Iowa. With exposed outside windows, the room was bright and inviting, one of their favorite spots. A wall separated the area from the rest of the basement. That afternoon they had heard the warnings about severe weather and the possibility of a tornado but knew they were in a safe place downstairs. Suddenly the television went off the air. The couple decided to move to the other side of the wall for added protection.

That's when Virgil heard a cry for help. A young mother and her two children, ages two and four, were outside, caught in the path of the storm. The family had been driving through town when they saw the monster approaching. "She took shelter in a car wash and realized that wasn't going to be enough protection," said Virgil. "That's when I heard her holler for help. She called, 'Can I come in? Can I come in?'" Virgil quickly shepherded the threesome into the basement. Within five minutes the tornado hit.

"It was a huge roar, like a freight train roar. We could see the ceiling of the basement rocking up and down, and then it was calm. I was going to get up and my wife said to wait a minute, and then we got it again. So we must have been in the eye of the tornado. Because it got quiet and then the storm punched again as loud and nasty as before."

Moments later, daylight could be seen at the front end of the basement. The house had been blown away and only the deck remained. "I found a board shoved into the wall like an arrow," remembered Virgil. "It was unbelievable." But they were all fine.

As Parkersburg's director of economic development, Virgil was swamped with the recovery efforts. It wasn't until days later that the shock of what had happened finally sank in. "My daughter came down from Minneapolis and my son from Arizona. When they saw the damage, well, that was hard," he said with a glint of tears in his eyes.

Goodrich plans to rebuild on the same lot in Parkersburg. Even with the house rebuilt, it will never be the same. Not only did the tornado steal their house, but it also pulled up and destroyed more than sixty trees that the couple had planted.

NINE LIVES AND COUNTING

The Sinclair grain elevator sits just a few miles east of Parkersburg. Built in 1979, Sinclair has a storage capacity of 2.3 million bushels and is considered one of the larger feed mills in the area. It is a major business for Parkersburg.

Kendall Vry is one of Sinclair's dozen or so employees. For the last few years he has punched in every weekend to grind feed for local livestock. During the week, he works on his own farm, growing corn, beans, and hay.

Shortly before 7:00 that Sunday morning, Vry checked into work for what would be a twelve-hour shift. The morning went smoothly, but by afternoon Vry noticed the weather beginning to turn ugly. He turned on the small television in his office and saw that a tornado warning had been issued. That sent Vry and another employee, Stan Mulder, outside to take a look. "We saw what appeared to be the beginning of a tornado. It wasn't turning and there was no tail to it whatsoever. It was a wide area of black," said Vry. The reports on television indicated the storm was tracking to the northeast of Sinclair, so Vry went back to work grinding feed. Minutes later, Mulder walked in with a silver dollar–sized chunk of hail. "He thought I had thrown it at him," said Vry. "We like to joke around with each other. But I told him I had nothing to do with that!"

Vry and Mulder hustled outside and watched the tornado enter the west side of Parkersburg, three miles away. "The weather reports kept saying it was going northeast but it never did. It kept getting closer. We could see the south side of that huge black wall and it looked kind of stationary. On the north side, clouds were getting sucked in," he said. The wind started to pick up, and a young man and his fiance pulled into the grain elevator. Could they stash their vehicle inside one of the sheds? Vry directed them to the largest shed, but the electricity failed and the giant doors wouldn't budge. He urged the young couple to take shelter at his parents' home a few miles away. With the tornado pushing over the hill, they took off along with Mulder, who wanted to check on his mother.

Vry walked quickly toward his office but stopped when he saw two more vehicles wedged between two sheds. Debris had started to fly, and it was apparent time was running out. Vry quickly motioned for whoever was sitting in the vehicles to come inside the large shed. They wouldn't budge. "I think they thought they were protected where they were. I went through the shed and out a side door and tried to get everyone inside," he said. As one woman climbed out of the truck, another came running around the shed, screaming. Her teenage son was in the office. "I said, 'We'll get your boy,'" said Vry. He quickly rounded up the teenager and hustled everyone into a room no bigger than eight feet by eight feet that was essentially a small building inside the large shed. The room contained the electrical breakers and water heater. The four people huddled together, ducked down, and put their arms around one another.

"We weren't in very long and the tornado hit. I could tell the big part of the shed was gone and blowing over our heads. Then the force of the tornado pushed our little building at a forty-five-degree angle and debris was flying all around. The water heater was now leaning against us and you just wondered how much more there was. We waited a minute and it grew quiet, and then we knew that we had made it," said Vry.

Peering through the cracks in the building, the small group could see light; gray insulation covered everything. When they crawled out, no one expected the sheer devastation surrounding them. "The storm threw semis full of feed all around us just like they were toys. Gas leaks were everywhere and giant tanks sprayed anhydrous ammonia into the air," he said. The women's vehicles were totaled. The huge grain bins were ripped apart, the entire operation destroyed.

"I had used up all my lives," said Vry with a smile, referring to his many close calls throughout his lifetime. "So I think the women saved me because I had to borrow someone else's."

Rest in Peace—New Hartford cemetery damage. COURTESY TOM WAGNER

CREDITS LEFT TO RIGHT: TIM KING, NATIONAL WEATHER SERVICE, NATIONAL WEATHER SERVICE, NATIONAL WEATHER SERVICE, NATIONAL WEATHER SERVICE

MEASURING GREATNESS

THE EF-SCALE
WHAT IS IT AND WHY WE SHOULD CARE

Whether it's the temperature, how much snow is on the ground, the amount of rain that falls, or the amount of moisture in the air, meteorologists love to measure things. Thus, it's only fitting that weather forecasters gauge the strength of tornadoes by measuring the power of their winds. The first to do this was a slight, primly dressed Japanese physicist named Tatsuya Fujita.

Mr. Tornado, as he was often called, introduced the Fujita Scale in a 1971 paper titled, "Proposed Characterization of Tornadoes and Hurricanes by Area and Intensity." Fujita revealed in the abstract his dreams and intentions of the F-Scale (F for Fujita). He wanted something that categorized each tornado by intensity and area. The scale he proposed was divided into six categories:

* F0 (Gale)
* F1 (Weak)
* F2 (Strong)
* F3 (Severe)
* F4 (Devastating)
* F5 (Incredible)

Fujita started with the belief that about 40 percent of all tornadoes were weak. From there he steadily raised intensity levels as wind speeds increased from strong, to severe, to devastating. His final classification came for tornadoes that were rarest of all, ones that occurred just a few times a decade, a one-in-a-thousand tornado he labeled the F5. Fujita stated that in an F5 tornado, "Incredible phenomena can occur." Houses would likely disintegrate and automobiles could be launched like missiles.

Fujita and his staff showed the value of the scale's application by surveying every tornado from the Super Outbreak of April 3–4, 1974. The F-Scale then became the norm to define every tornado that has since occurred in the United States. The F-Scale also became the heart of the tornado database that contains a record of every tornado in the United States since 1950.

Over the years and for a number of reasons, flaws in the F-Scale were noted. One of the most obvious and troublesome was the fact that wind speeds were often overestimated in tornadoes greater than F3. Fujita himself noticed the need for revisions. A committee developed the Enhanced Fujita Scale, and a particular point was emphasized to board members: "It must continue to support and maintain the original tornado database." In other words, there must be some conformity to that of the original F-Scale. In 2007 a new scale was implemented, known as the EF-Scale (E standing for enhanced).

The new EF-Scale incorporates more damage indicators and degrees of damage than the original F-Scale, allowing more detailed analysis and better correlation between damage and wind speed. The original F-Scale historical database from 1950 remains unchanged. This means an F5 tornado rated years ago is still an F5, but the wind speed associated with the tornado may have been somewhat less than previously estimated. In a nutshell, the biggest difference in the new EF-Scale is that lower wind speeds are sufficient for higher end EF tornadoes.

ENHANCED FUJITA SCALE (EF-SCALE)

SCALE	WIND ESTIMATE *** (MPH) 3-Second Gust
EF0	65–85
EF1	86–110
EF2	111–135
EF3	136–165
EF4	166–200
EF5	Over 200

*** IMPORTANT NOTE ABOUT ENHANCED F-SCALE WINDS: The Enhanced F-Scale still is a set of wind estimates (not measurements) based on damage. It uses three-second gusts estimated at the point of damage based on a judgment of eight levels of damage and twenty-eight indicators. These estimates vary with height and exposure.

Important: The three-second gust is not the same wind as in standard surface observations. Standard measurements are taken by weather stations in open exposures, using a directly measured "one-minute mile" speed.

RATING THE TORNADO

Donna Dubberke, the National Weather Service warning coordinator based in the Quad Cities office, described her tornado fieldwork: "The Des Moines office had already done a preliminary damage assessment, standard practice within the Weather Service when we suspect damage is greater than EF3. The purpose is to get more experience and add additional opinions to see if everyone comes to the same conclusion. It was clear this was going to be bigger than a three. When the call came, I was actually out surveying near Hazelton. I decided to go to Parkersburg because this was such a rare storm and also to gain professional experience.

"Our focus when doing a survey is to document the tornado's beginning, ending, path width, and damage. Our focus was on the heaviest damaged areas in Parkersburg and near the cemetery in New Hartford where a woman was killed. We also looked at what wasn't damaged, because that can also be an indicator.

"The big challenge was that there was so much debris. In order to really assess the damage, you need to know where the debris came from. Once the tornado hits structures and picks up debris, it becomes debris loaded, and at that point the debris itself creates even more damage. When all the houses are strewn together, the questions become, whose debris is it and where did it come from?

"The keys to establishing an EF5 rating are found in the details. When you get into a five status, there are not a lot of indicators left to search. Not many structures can withstand winds up to 200 miles per hour. For example, trees don't make it.

"There were some EF5 indicators in the homes right by the golf course. These were newer homes, pretty well connected to their foundations, that were completely demolished, and debris pushed off the foundation altogether. In the areas where the damage was severe and where there were some fatalities, a gentleman we talked to had a large van that had been lifted and carried several blocks away, and that was an indicator.

"Another indicator was a number of homes that were completely gone, nothing left. Another one was the way in which the remaining debris was broken into such small pieces. We didn't just see a broken two-by-four, we saw small pieces of splintered wood. Another thing for me was the duration. Another gentleman said the tornado was thirty to forty-five seconds tops in terms of duration, and to produce that kind of damage in that amount of time helped convince me to push it to the top of the EF-Scale.

"Additionally, there was another spot on the southwest edge of town where some type of metal industrial building was being converted into a church. It contained some metal I-beams in the construction and those I-beams were sheared off at the ground.

"Another problem we encountered was our inability to speak with people in spots with severe damage. Unfortunately, because of the fatalities

No smiles for the camera—documenting the carnage.
COURTESY LENA EGGERS AND DON RADLOFF

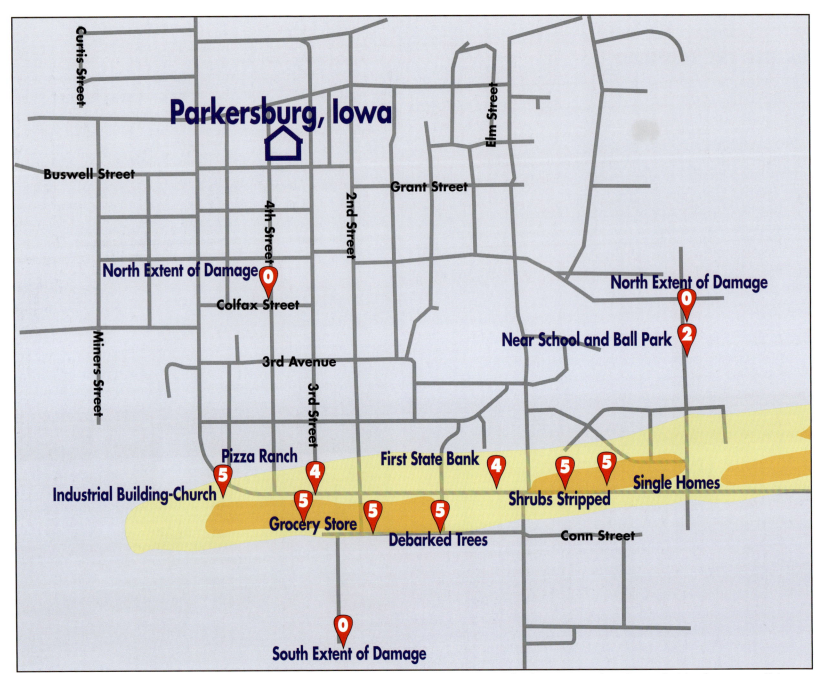

This damage survey by the National Weather Service shows that the most severe destruction in Parkersburg occurred on the south side of town, parallel to Highway 57. In this two-block corridor, EF4 and EF5 damage (winds of 166 to 205 mph) were common. COURTESY NATIONAL WEATHER SERVICE DSM

and injuries, people are not there to talk to you or they are not able, they are tired, they have been up for all hours, it is a very difficult time.

"In the end, it is important to get multiple indicators in the same spot to make a determination on the wind speed. So we took all the indicators we felt were over EF5 and, while we didn't feel we could go as strong as 210, we did come in at 205 miles per hour.

"When you look down over the valley, there is just no way to capture that devastation on film. Standing in the middle of it, looking at the scope of it and seeing the look on people's faces, I was trying to figure out how in the world you can describe that to somebody. It is truly overwhelming. It is almost surreal. There is so much activity going on all around you. Small children walking around in the debris, and you are worried they are going to step on nails. I was okay, you have to detach yourself from it and focus on your job.

"When we complete our survey, we usually put out a local storm report with our rating. In this case, the first thing we did was let the local officials know so they didn't hear it in a roundabout way. Then there was a briefing to the state. Des Moines then files an official report and lets our region know and our headquarters know, because it is so rare. Then it goes into storm data, where it becomes a permanent record for the weather world."

SIGNIFICANT EF5 IOWA TORNADOES AND THEIR COMPARISON TO THE PARKERSBURG STORM

Twelve EF5 storms are now part of Iowa's elite tornado history. Dating back to 1860, many of the state's twisters pre-date present-day standards regarding tornado forecasting, public warning, and education. The Weather Service set these standards in the 1950s. Any Iowa tornado since 1950 is considered to have occurred during the modern-day era of tornado classification.

The following chronological examination of EF5 tornadoes and their similarity to Parkersburg is broken into two categories: one comparing Parkersburg to all tornadoes and one comparing Parkersburg to modern-day twisters from 1950 to the present.

The Parkersburg tornado was one for the ages. Those who saw and lived "the funnel" speak reverently about it, the shock and awe of the experience tucked deep inside. Where does this storm stand when compared to the other EF5 giants of Iowa's tornado past?

To get the answer, start at the beginning. Tornadoes were part of Iowa's natural environment long before any settlers arrived on the scene. The state's earliest twisters played to an audience of prairie grass, wildlife, and Native Americans who left no written accounts.

As explorers arrived in the state, so, too, did reports of tornadoes, the earliest logged July 29, 1804, when the Lewis and Clark Expedition paddled up the Missouri River along Iowa's southwestern border. There they observed the destructive work of a tornado that had passed through Iowa

The true meaning of "compact car"—near New Hartford.
COURTESY JUSTIN STOCKDALE

MONTHLY WEATHER REVIEW

NO.	LOCATION	DATE	TIME OF OCCURRENCE	KILLED	INJURED
1	Near Iowa City	May 24, 1859	Aftn.	5	18
2	Camanche	June 3, 1860	6:30 p.	141	329
3	Keokuk and Washington Co.	May 22, 1873	2:00 p.	8	15
4	Monona Co. to Buena Vista Co.	April 21, 1878	4:00 p.	10	28
5	Crawford Co. to Pocahontas Co.	April 21, 1878	4:40 p.	18	29
6	Montello	Oct. 8, 1878	5:30 p.	0	0
7	Macedonia	June 9, 1880	Evening	Many	----------
8	Grinnell	June 17, 1882	8:45 p.	100	----------
9	Coon Rapids	April 14, 1886	3:00 p.	Destroyed much of Coon Rapids	
10	Pomeroy	July 6, 1893	5:00 p.	89	200
11	Kossuth and Clay Co.	Sept. 21, 1894	5:00 p.	53	100
12	Sioux Co.	May 3, 1895	3:20 p.	15	35
13	Polk Co. to Jasper Co.	May 24, 1896	9:30 p.	20	----------
14	Stanwood	May 18, 1898	3:00 p.	19	40
15	Harlan	March 23, 1913	6:00 p.	33	100
16	Pearl Rock to Calmar	May 9, 1918	4:00 p.	8	20
17	Crawford and Carroll Co.	May 21, 1918	2:15 p.	6	35
18	Berkley to Wellsburg	May 21, 1918	3:45 p.	10	91

Eighteen significant Iowa tornadoes prior to 1920.

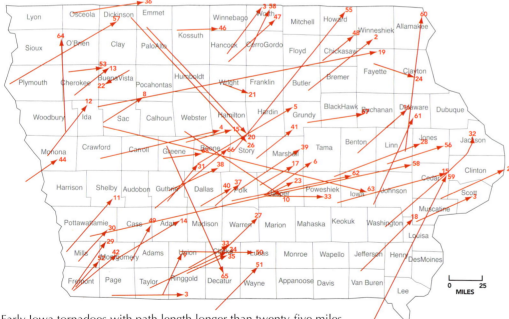

Early Iowa tornadoes with path length longer than twenty-five miles, 1916–1960. COURTESY IOWA MONTHLY WEATHER REVIEW 1962

one year earlier. Captain Clark wrote in his journal: "On the S.S. passed much fallen timber apparently the ravages of a Dreddful harican which had passed obliquely across the river from NW to SE about 12 months Sine, many trees were broken off near the ground the trunks of which were four feet in diameter."

Iowa's written tornado history began to flourish with the signing of the Black Hawk Treaty in 1833, which allowed the population to rapidly increase. With more people came more reports. Between 1837 and 1887, Lt. John Finley of the U.S. Signal Corps compiled a list of 120 documented twisters in his book *Iowa Tornadoes*.

In June 1859, J. A. Wetherby of Iowa City wrote about a tornado that passed south of the city on the afternoon of May 24, 1859. According to the editor of the *Iowa City Republican*, loss of life and property was heavy, with five killed and eighteen injured. Dr. Gustavus Hinrichs, the first director of the Iowa Weather Service, authenticated this storm in his journal as, "The first reported tornado in the State after settlement began, though undoubtedly many must have occurred previously that were not reported."

The following summer, on June 3, 1860, a tornado obliterated the town of Camanche in east-central Iowa. *Harper's Weekly* out of New York reported the tornado as a "national calamity" to its readers around the country. In turn, the awareness of tornadoes in the United States was greatly heightened. More important to Iowa, the Camanche storm became the first documented EF5 tornado in state history.

This, then, is where the story of Iowa's most celebrated tornadoes begins. In chronological order, here are historical accounts of all the EF5 Iowa tornadoes that preceded Parkersburg.

EF5 IOWA TORNADOES PRIOR TO THE MODERN ERA (1860 TO 1950)

THE CAMANCHE TORNADO
Date: June 3, 1860
Deaths: 141
Injuries: 329
Counties: Hardin, Grundy, Tama, Benton, Linn, Cedar, Clinton
Path Length: Unknown, possibly in excess of 200 miles

The Camanche tornado of 1860 was dangerous, unannounced, and deadly. On the ground for hours, it traveled from Hardin County in central Iowa through Cedar Rapids, paralleling the old railway line that is now U.S. Highway 30, until it smashed into the unsuspecting village of Camanche. In a few short minutes, the town of 1,200 founded two years earlier was destroyed. Nearly every citizen was injured and many were killed. From here the violent storm crossed the Mississippi River, leveled Albany, Illinois, and dissipated over Lake Michigan just north of Chicago.

Traveling at speeds estimated up to 66 miles per hour, the tornado left 141 dead and another 329 injured as it raked a trail of devastation at least half a mile wide. It's reported that many of the injured died later, pushing the unofficial death toll even higher. While there is no official EF-Scale designation to the Camanche storm, based on damage accounts, duration, width, injuries, and fatalities, it seems especially deserving of an EF5 rating. According to historian Benjamin Gue, the lieutenant governor of Iowa from 1866 to 1868, "several meteorologists made careful investigation to ascertain the velocity of the circular motion of the winds

This *Harper's Weekly* rendering illustrates the Camanche tornado's devastation: only twenty buildings were still on their foundations.
COURTESY CAMANCHE HISTORICAL SOCIETY

Located in Rose Hill Cemetery in Camanche is a small plaque mounted on a rock that says, "Sacred to the memory of those who perished in the Camanche Tornado, June 3, 1860." COURTESY JEFF HANSON

A *Harper's Weekly* artist drew this rendering of the coffins of Camanche residents killed by the 1860 tornado.
COURTESY CAMANCHE HISTORICAL SOCIETY

which wrought such fearful destruction, that it was at the rate of 300 miles per hour."

A correspondent for the New York Herald happened to be in Camanche at the time of the storm and two days later gave this unique perspective from Clinton, Iowa: "The air, darkened by the immense moving cloud, charged with death, the rain, which was now falling in torrents, the fragments of crushed and scattered buildings, which were flying all directions, and the shrieks and groans and prayers for help were heard, even above the din and roar of the tempest, all combined, rendered the scene one of the most solemn and extraordinary I ever witnessed . . . the angel of destruction passed over it, and with his wings had brushed it from the bosom of the plain."

Amid the shattered surroundings following the storm, survivors in Camanche held funerals for lost loved ones. Two days later, twenty-five coffins were neatly arranged for a mass funeral in front of what had been Dunning's Bank. The History of Clinton County contains this sobering account: "Over two thousand sympathizing friends and neighbors were present and frequent outbursts of grief amid the deep hush that pervaded the assemblage attested to the profound grief of the stalwart men as well as the tender-hearted women."

Old-timers would tell about the storm for many years to come. But as they passed away, so, too, did many of the astonishing tales. But even today, the Camanche tornado, as it became known, lives on as not only one of the most severe Iowa tornadoes but as one of the worst ever experienced by the settlers of the region known as the Old Northwest.

THE GRINNELL TORNADO

Date: June 17, 1882
Deaths: 100
Injuries: 300
Counties: Green, Boone, Story, Marshall, Jasper, Poweshiek
Path Length: 105 miles

"June 22, 1882—On last Saturday evening at 8:45 P.M. the most terrible tornado ever known in Iowa and perhaps in the United States, passed through central Iowa destroying part of Grinnell and Malcom, both in

The Great Grinnell Cyclone—1882.
COURTESY STEWART LIBRARY ARCHIVES, GRINNELL, IOWA.

Poweshiek County. We give here a complete report condensed from the accounts given of the cyclone in the Grinnell Herald:

"Great Grinnell Cyclone–1882. The most terrific disaster in the history of Iowa is the one of which our now desolate town at Grinnell is the victim. The peculiar aspect of the sky was matter of common remark on the streets of yesterday afternoon. An hour or more before sunset the northern sky was hung with conical, downward-pointing clouds, the like of which none of us had ever seen. After sunset, and even after darkness was gathering, the western sky half way to the zenith was lurid and brilliant and unearthly—an ominous sight which fascinated while it filled one with ill-dread. Almost ere the brilliant apparition in the west had disappeared the storm broke. It was accompanied with a roaring like thunder, or perhaps more like rumbling of a dozen heavy freight trains. Chimneys, trees, houses, barns began to fly like leaves. People took to their cellars."

This 1882 account from the Republican, a Keosauqua, Iowa, newspaper, is clear proof that tornadoes have been a source of fear and fascination in Iowa as far back as written state history goes. One hundred people were killed in this 105-mile-long EF5 twister, which settlers of

Remains of a piano after the 1882 cyclone.
COURTESY STEWART LIBRARY ARCHIVES, GRINNELL, IOWA.

FIG. IV.—AS IT STRUCK POMEROY.

Appearance of the storm cloud as it struck Pomeroy. In *Story of a Storm* by F. W. Sprague, 1893. COURTESY NOAA

the day stated was the worst tornado ever known in Iowa. While the violence of the storm is not disputed, where the Grinnell tornado stands in relation to Iowa EF5 history is much harder to gauge.

It's believed the Camanche tornado, which hit the state twenty-two years earlier, actually set the benchmark for all tornadoes to come. Apparently due to the sparse population of Iowa and the word-of-mouth nature of conveying news, many of the Grinnell survivors were unaware of this earlier EF5 tornado that had passed through the eastern half of the state.

THE POMEROY TORNADO

Date:	July 16, 1893
Deaths:	89
Injuries:	200
Counties:	Cherokee, Buena Vista, Pocahontas, Calhoun
Path Length:	55 miles

One of the state's deadliest twisters, this EF5 storm developed relatively late in the season. The tornado originated in the northwestern Iowa county of Cherokee around 5:00 P.M. and moved east-southeast before arriving in Pomeroy between 6:30 and 7:00 P.M. The tornado reportedly bounded along the prairie like a huge ball as it swept away homes in four different counties. In rural Cherokee County, six people in a single family were killed. Chickens were reportedly found alive but completely stripped of all feathers.

J. R. Sage of the Iowa Weather and Crop Report Service noted that many lives were spared in Pomeroy because people took refuge in tornado caves and basements. The report further stated: "If every family had been provided with such a place of refuge, and had heeded nature's danger signals, no fatalities would have been recorded. Their cost is trifling, and at such a time a very poor hole in the ground is worth more to a family than the richest gold mine in America."

In the town of Pomeroy, 80 percent of the homes were severely damaged and forty-nine people perished. In *History of Iowa*, Benjamin Gue wrote this account of what Pomeroy physician Dr. D. J. Townsend said about victims he treated: "The wounds were not of a class that were met with in any other calamity than a tornado. The tissues were bruised,

A young boy with a calf, the sole survivors of one farm struck by the tornado. COURTESY NOAA

Soldiers patrol in the ruins of Pomeroy. COURTESY NOAA

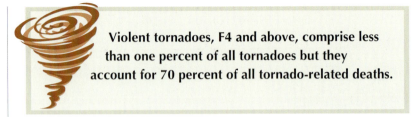

Violent tornadoes, F4 and above, comprise less than one percent of all tornadoes but they account for 70 percent of all tornado-related deaths.

punctured, incised, and lacerated with the addition of having foreign matter of every conceivable kind literally ground into the flesh." He went on to say: "Many were contused from one end of the body to the other. The dirt and sand plastered upon and into the skin in such a manner that it was extremely difficult to remove them."

In a 1962 report on Iowa tornadoes, the *Iowa Weather Review* mentions another intriguing aspect of the storm: "This tornado also passed over Storm Lake creating a waterspout and the water along the north shore receded rapidly, a hundred feet or more, leaving bare ground at the pier where the small steamboat lands. After its passage, the water rushed back with a tidal wave several feet high."

THE KOSSUTH AND CLAY COUNTY TORNADO

Date:	September 21, 1894
Deaths:	53
Injuries:	100
Counties:	Kossuth, Hancock, Winnebago
Path Length:	50 miles

The worst of this fifty-mile-long EF5 twister occurred in Kossuth County, where thirteen people lost their lives. The storm originally set down around 5:00 P.M. in an open corn field. According to the *Iowa Weather and Crop Report* from 1894, a resident of the area named Peters "claims to have distinctly seen three of the death dealing monsters; one was to the north, the other to the south of him and had passed east of his place, moving in a zigzag course; the third one, coming from the west, was almost upon him when he discovered it; he barely had time to drop into his cyclone cellar when it passed over, demolishing his house and out buildings."

THE SIOUX COUNTY TORNADO

Date: May 3, 1895
Deaths: 15
Injuries: 35
Counties: Sioux
Path Length: 13 miles

This tornado had a relatively short life span but made up for it by producing nine fatalities and thirty-five injuries as it advanced north-northeast through Sioux County. According to the *Iowa Weather and Crop Report* from 1895, "Immense hailstones fell on a track four or five miles to the left of the tornado."

One of the more tragic aspects of this storm was that it flattened four rural schools in the area around Ireton and Hull. According to the book *Significant Tornadoes* by Tom Grazulis, "Two school houses several miles apart were leveled killing teachers and students. The dead teacher at the first school (George Marsden at the Haggie School) was the brother of the teacher killed at the second school (Ann Marsden at the Coombs School). Adjoining farms were entirely leveled, with several deaths in homes. School children were carried for up to a half mile, and many sustained injuries that would be life-long burdens. Some publications put the death toll at fifteen."

This EF5 twister scorched a path of destruction thirteen miles long.

THE CRAWFORD AND CARROLL COUNTY TORNADO

Date: May 21, 1918
Deaths: 6
Injuries: 35
Counties: Crawford, Carroll, Greene
Path Length: 37 miles

For the most part, this EF5 tornado was a rural event, with the twister tracking from near Denison to the town of Stanhope. Even so, with a half-mile-wide damage path, it laid waste to at least twenty farmsteads and carried debris miles away. One couple trying to outrace the storm in a buggy was overtaken and killed near the town of Churdan. All told, the tornado killed four people and injured thirty.

EF5 IOWA TORNADOES (1950 TO THE PRESENT)

THE ADAIR TORNADO

Date: June 27, 1953
Deaths: 1
Injuries: 2
Counties: Cass, Adair
Path Length: 10 miles

Little is known about this major tornado, due in large part to the fact that it was of short duration and essentially a rural storm. The tornado formed five miles southwest of Adair and pounded a trail ten miles long before it abruptly dissipated southeast of town. The twister mangled four farms along its way. As it reached EF5 status along the Cass-Adair county line, it produced damage so severe at one farm there was virtually nothing left at the site. Heavy machinery was reportedly thrown over a hundred yards, and boards were found driven into trees.

Traveling over a remote population base and with a short life span, the tornado injured only two people. One person was killed just south of Adair. According to state figures, the tornado was the first EF5 recorded in Iowa since May 1918.

THE BELMOND TORNADO

Date: October 14, 1966
Deaths: 6
Injuries: 172
Counties: Wright
Path Length: 20 miles

Around the state of Iowa, a typical mid-October Friday culminates with a football game at the local high school. By then the leaves are flashing crimson, the days are crisp, and the nights are often cold. The season's first snow is by far a more likely topic of conversation than tornadoes and severe thunderstorms.

In terms of weather, October 14, 1966, broke the mold. That

COURTESY *DES MOINES REGISTER*

Open air market—Belmond store following the storm.
COURTESY BELMOND HISTORICAL SOCIETY

It's not known if the above tornado eventually became the Belmond twister. However, it is known to be one of several rare fall tornadoes that touched down that afternoon. COURTESY NATIONAL WEATHER SERVICE

morning the weather was balmy by October standards and, most notably, unusually moist. Outside of Belmond, fields of corn, golden and ready to harvest, swayed in the warm southerly winds. Excitement was in the air as this Friday was homecoming, and the Belmond Broncos' thirty-game winning streak was on the line against archrival Lake Mills. A midday parade was planned to commemorate the event.

By early afternoon, the townspeople and students of Belmond converged on Main Street for the start of the festivities. With organizers unaware of any tornado threat, the homecoming parade went off without a hitch. An unwelcome participant was a potent cold front approaching from the west. It would soon be the catalyst for an epic fall tornado that would crash the party.

Following the event, students crammed into their local hangout, the Townhouse, for something to eat and drink. Outside, many of the townspeople lingered on the street. Then word arrived of a dangerous storm headed north out of Clarion that may have been producing a tornado. Within minutes the street was cleared, and Belmond resembled a ghost town while townspeople awaited its fate.

At 2:55 P.M. the power failed in Belmond and the EF5 twister barged into town. The windows in the Townhouse exploded, and, within seconds, the landscape of Belmond was changed forever. Six residents were killed and 172 sustained injuries. Seventy-five of the town's 112 businesses were destroyed or damaged.

One Belmond woman found, amid the rubble of her home, that a bed was still perfectly made, except for a lump under the bedspread—which turned out to be two pieces of artificial fruit that had been in a dish on her dining room table. "The dish never did come home," she said in Belmond's *Tornado Book*, a collection of stories and pictures commemorating that horrifying day.

By the end of the evening, twelve tornadoes had skipped across Iowa. As Belmond picked up the pieces, four inches of snow fell just to the west in Sioux City.

Being an EF5 storm, the Belmond tornado was a rarity unto itself. Coming in mid-October makes it exceptional and highly unique when compared to Iowa's other strong tornadoes.

THE CHARLES CITY AND OELWEIN TORNADOES

THE CHARLES CITY TORNADO

Date:	May 15, 1968
Deaths:	13
Injuries:	450
Counties:	Franklin, Butler, Floyd, Chickasaw, Howard
Path Length:	62 miles

Two of Iowa's worst modern-day tornadoes swept through the state within seven minutes of one another on May 15, 1968. Spawned by an anomalously deep low pressure center passing through northwest Iowa, both tornadoes achieved EF5 status. Including the Parkersburg tornado, Iowa has experienced only six EF5 twisters since 1950, which highlights the exceptional nature of having two occur within fifty miles of one another at virtually the same time.

As with the Parkersburg tornado forty years later, the afternoon preceding the Charles City storm was unseasonably warm and muggy. Despite the volatile meteorological conditions, there had been no signs of serious weather trouble until 4:10 P.M., when the first tornado of the day set down

Wedge tornado bears down on Charles City.
COURTESY FLOYD COUNTY HISTORICAL SOCIETY

on the edge of Aredale, about twenty-five miles southwest of Charles City. As the storm steamed northeast, it produced multiple tornado sightings while it skirted Marble Rock and closed in on Charles City.

Pressing on, the tornado was soon visible outside Charles City, growing wider and more intense. Its course would just graze or perhaps even miss the southeast corner of town. However, in a tragic twist at 4:50 P.M., it abruptly turned and roared directly north through the heart of Charles City.

While there was some warning, only a few residents ever clearly saw the huge tornado before it barreled into town. Twelve-year-old Jeff Sisson got a quick glimpse and described it: "When my father pulled in the driveway, we stood there for just a few minutes looking at the sky. It was getting black in the southwest. I remember looking straight up in the air and seeing these really strange clouds bulging down toward the ground. It's something I have seen since, but never quite the way these looked.

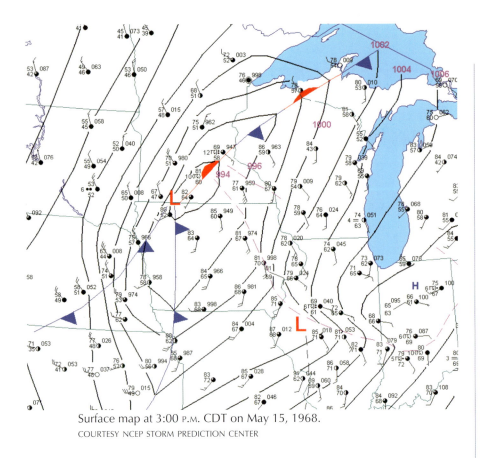

Surface map at 3:00 P.M. CDT on May 15, 1968.
COURTESY NCEP STORM PREDICTION CENTER

The north end of Charles City just a few blocks west of the Oliver Plant.
COURTESY JEFF SISSEN

McKinley School building. COURTESY JEFF SISSEN

Just a minute or two later, about 4:45 P.M., it hailed. The size was enormous. I remember them splatting on our driveway and thundering off of my father's International Scout. We ran up on the covered front porch, which faced due west down Third Avenue.

"My father and I noticed the clouds twisting, rotating from the southeast to the northwest. It was amazing to watch, because there was very little breeze on the ground but it still looked so violent. As we looked to the west, above the treetops we could see stuff flying around in the air. I thought it was leaves and sticks, but realized later it was huge chunks of debris. The wind then came up and blew very hard, and we headed to the basement. I looked out a window that faced north and watched the trees move violently. The winds lasted a couple of minutes and then stopped."

After emerging from the basement, Jeff and his family headed for a part of town that had taken a punishing hit. Jeff continued, "When we reached the corner of Third and Grand Avenue, the steeple of the Immaculate Conception Catholic Church had crashed down through the roof and left a huge hole. We couldn't believe our eyes as we looked up at the damage . . . bricks strewn all over the front steps and street. It was about this time when everyone in our vehicle looked west toward downtown Charles City, and we were aghast. It looked like a bomb had hit. All I remember is seeing horrible destruction, people wandering in shock, and streets clogged with debris, live wires, and silence . . . this huge mess and no noise. It was truly unbelievable."

In the unimaginable aftermath, 337 homes and fifty-eight businesses lay victim to the tornado. Eight churches and three schools were damaged or destroyed, the police station was heavily damaged, and 1,250 vehicles were destroyed. Thirteen people were killed and 450 others sustained injuries. The thirteen deaths were the most in an Iowa tornado since a north-central Iowa twister killed fourteen in September 1894.

THE OELWEIN TORNADO	
Date:	May 15, 1968
Deaths:	5
Injuries:	156
Counties:	Fayette
Path Length:	13 miles

A forty-five-minute drive from Charles City, in the rolling hills of northeast Iowa, the community of Oelwein braced for its own date with disaster. It was here, minutes after the destruction of Charles City, that another twister (perhaps two) swooped from the sky one mile southwest of town. The tornado intensified at a frenetic pace and spun into an EF5 in the few yards left between it and Oelwein.

With 1968 radar technology so basic and the storm evolving so quickly, the twister was on top of Oelwein before most knew it even existed. Warning sirens blew just fifteen seconds before the power was cut. Then, like a traveler passing through town, the tornado drove right up the main business street and in a matter of three or four minutes was gone.

Testifying to the lack of warning and prevailing shock in Oelwein, Fire Chief Wally Rundle gave this account in a 2006 interview with Roger King of radio station KOEL: "I was at the firehouse on the late afternoon of May 15, 1968, when the tornado struck. There was no warning that a devastating storm was coming. It was actually two tornadoes that started south of Oelwein and then merged into one and headed into Oelwein. I started to see hail come down, and the next thing I knew people were running past the station looking over their shoulders. One woman died in the storm. We rescued the woman from an apartment in the downtown area, as she had fallen down between two floors when the chimney blew out. The woman was taken to the Oelwein hospital where she later died from her injuries. The storm caused massive damage to the downtown and eastern side of the city."

Meantime, at radio station KOEL, engineer Dean Meyer was broadcasting a tornado warning when the funnel hit. He cried, "God help me," as the power was cut and the station tower crashed to the ground. Meyer escaped unharmed.

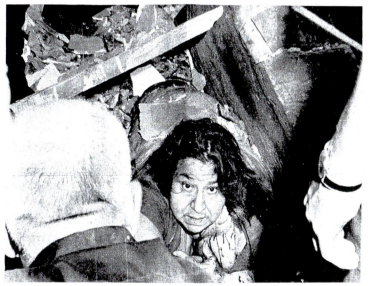

Oelwein tornado victim. COURTESY OELWEIN DAILY REGISTER

The post-storm wrap-up showed five people perished in the Oelwein storm and 156 were injured. Sixty-eight homes were destroyed and 132 sustained major damage. Every business in the district suffered damage; fifty-one were destroyed. Two churches, an elementary school, and the middle school were also destroyed. In addition, after leaving Oelwein, the storm leveled five square blocks in Maynard.

THE JORDAN TORNADO

Date:	June 13, 1976
Deaths:	0
Injuries:	9
Counties:	Boone, Story
Path Length:	21 miles

With only a handful of EF5 tornadoes in Iowa since 1950, it's common knowledge that storms of such magnitude are few and far between. What made the Jordan EF5 exceptionally unusual was its violent nature and the fact that it was anti-cyclonic.

Anti-cyclonic Jordan tornado east of Boone. COURTESY H. C. ANDERSON

Prior to the Jordan storm, science-based belief suggested that all tornadoes in the northern hemisphere rotated in a counter-clockwise direction. This notion gave rise to early American Midwest settlers using the term cyclone to refer to the violent columns of wind that swept the prairies. So popular was the term cyclone that, in 1900, author John Baum used it in his book, *The Wizard of Oz*, to describe the fearsome tornado that swept away Dorothy.

The cyclonic nature of tornadoes remained unchallenged for decades. Then, on a warm sticky day in June, a massive tornado developed near Ames, Iowa. WHO-TV photojournalist Charles Barthold captured the monster on video; while his footage nearly got him killed, it set the meteorological world on its ear.

His extraordinary video soon landed in the hands of world-renowned tornado expert Dr. Ted Fujita, creator of the F-Scale used to rate tornado intensity. After intensive study of the footage, Fujita claimed the twister was the most violent he had ever seen. Upon further review he also noticed the tornado was rotating clockwise, making the Jordan storm the first known anti-cyclonic tornado in the world. Meteorologists have since documented other anti-cyclonic tornadoes, but they account for only a select few of those that ever touch ground.

As for Barthold, just getting the historic video was a harrowing experience. In 1976 storm chasing was in its infancy; mobile technology utilizing radar and GPS positioning was nonexistent. Alone in his car, Barthold stalked the storm near a cornfield where the twister loomed. Because the tornado was wrapped in rain, Barthold was unaware of the tornado's exact position. He decided to make a move toward the storm, drove through the rain, and suddenly was within a hundred yards of the tornado, a dangerous maneuver that chasers call core punching. The car windows popped out, and showers of rock and debris battered his car. It was nearly a fatal mistake, but he survived his encounter with an EF5 tornado.

The reward for his work with a Bell and Howell movie camera was a Peabody Award for TV News. At the time he was the youngest ever to win the coveted honor.

As for the tornado, its unusual nature was highlighted by a 110-degree turn that resulted in a rare U-shaped path. For most of its cycle the twister was over open country, so most of the significant damage occurred on farms in and around the hamlet of Jordan, which itself suffered extensive damage. The rural nature of the storm meant no fatalities and only nine injuries, another rarity for an EF5 storm.

PARKERSBURG'S RANKING COMPARED TO ALL OF IOWA'S EF5 TORNADOES

It's difficult to compare the Camanche, Grinnell, and Pomeroy tornadoes to the Parkersburg storm because they occurred over a century ago. Documentation and knowledge of tornadoes in the 1800s was very limited, and construction methods and materials have changed greatly. Since there was no advance warning with any of the older storms, injecting today's warning and communication system into the events would have saved untold lives.

Speculation, however, does not alter the hard facts. In terms of fatalities, the Camanche storm produced substantially more deaths than any other tornado in state history (133 more than Parkersburg). Combine the twister's death toll with its long track and wide damage path, and a strong case unfolds for ranking the Camanche tornado as the worst to ever strike in Iowa's written history. Using the same criteria, the Grinnell tornado with its 100 fatalities makes it a close runner-up. Best estimates on the Parkersburg storm would have it challenging for third place on the all-time worst list, along with the Charles City and Pomeroy tornadoes.

PARKERSBURG'S RANKING AGAINST ALL MODERN-DAY IOWA TORNADOES (1950 TO PRESENT)

Of all Iowa's tornadoes since 1900, only the modern-day Charles City twister appears on a par with the Parkersburg storm. In many ways, the two were eerily similar. Although hitting in different years, both tornadoes struck within nine minutes of one another on dates just ten days apart. Each produced a swath of destruction half to three-quarters of a mile wide. Both were long-tracked wedge tornadoes, with path lengths exceeding forty miles. Each destroyed approximately 350 homes, and about 50 percent of each city was damaged. Finally, each tornado originated within thirty miles of one another in Butler County, and grew to be EF5s with winds exceeding 205 miles per hour. Debris from both storms was found up to eighty miles away in the adjoining states of Wisconsin and Minnesota.

While the death toll was nearly twice as high in Charles City (eight in Parkersburg/New Hartford vs. thirteen in Charles City), this was probably due more to advanced warning than the intensity of the individual tornadoes. Thanks to high-resolution Doppler radars at National Weather Service offices (unavailable in 1968), tornado warnings were issued for Parkersburg forty minutes before the tornado arrived. Add to that the sophisticated warning systems at local television stations, which displayed the warnings within seconds of their issuance. That, combined with sirens, spotters, cell phones, weather radio, the Internet, and overall improved communication gave Parkersburg advance warning that Charles City never had.

It is reasonable to conclude that today's technology saved countless lives and injuries in Parkersburg and surrounding areas. Therefore, there is no clear answer when considering the question of which was Iowa's worst modern-day tornado, Parkersburg or Charles City. Both were heavyweights that dealt knock-out blows! ✸

Parkersburg enters the arms of an EF5 tornado. COURTESY CHRIS BIRCH

CREDITS LEFT TO RIGHT: TIM AND KYLE WOLTHOFF, ASHLEY PAVELEC, DAN BLEVINS, DAN BLEVINS, DAN BLEVINS

WORKS of the DEVIL

OR ACTS OF GOD
A HISTORY OF TORNADO FORECASTING

Forecasting tornadoes has had varying levels of success throughout the years. The National Weather Service and its predecessors have predicted and warned communities of these severe weather threats with ever-increasing accuracy that has saved countless lives and billions of dollars. For the 200th anniversary of the National Oceanic and Atmospheric Administration (NOAA) in 2007, NOAA forecasters and scientists prepared the following account of the history of tornado forecasting.

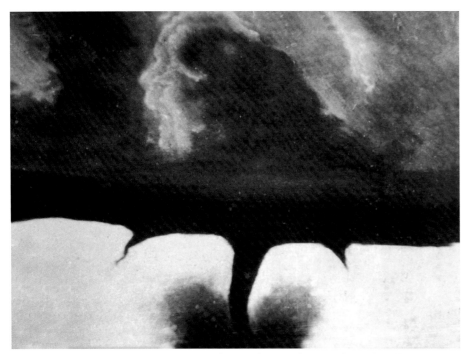

August 28, 1884, twenty miles southwest of Howard, South Dakota.
COURTESY NOAA, NWS COLLECTION

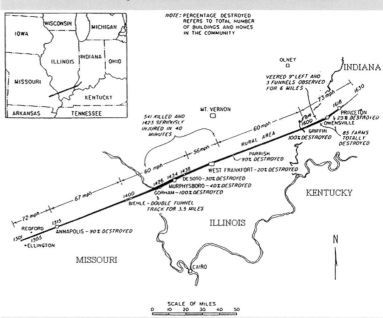

The Tri-State Tornado Track, courtesy "Illinois Tornadoes" by John W. Wilson and Stanley A. Changnon, Jr., Illinois State Water Survey, Urbana, IL (1971).

Imagine early explorers of the New World, traveling westward across America and encountering many new sights and experiences. These explorers would see and hear thunderstorms and experience hail, strong hurricane-force winds, and tornadoes for the very first time. Did they think these acts of nature were works of the devil or acts of God?

The first possible tornado report in the United States occurred in July 1643 in Lynn, Newbury, and Hampton, Massachusetts, documented by author David Ludlam. The report was recorded by Massachusetts governor and weather enthusiast John Winthrop, who observed a sudden gust that whipped up dust, lifted his meeting house, and killed a nearby observer with a fallen tree.

EARLY ADVANCES

In 1882, after nearly 300 years of numerous observations and stories of whirlwinds, cyclones, and tornadoes, U.S. Army Signal Corps Sergeant John P. Finley was placed in charge of the investigation of tornadoes and the development of forecasting methods. During a tornado outbreak that occurred on February 19, 1884, Finley established 15 rules for early tornado forecasting. In 1888 Finley published these rules, which identified signs that the formation of a tornado was likely.

Unfortunately, as Finley was developing his techniques, the tornado prediction program encountered a huge roadblock. The word "tornado" was banned from official forecasts by the U.S. Army Signal Corps due to limitations with the observing network and concerns over causing mass panic among the general public.

THE TRI-STATE TORNADO

The ban on the use of the word "tornado" continued for four decades, into the early twentieth century. During that time one of the biggest, longest-tracking tornadoes occurred. Known as the Tri-State Tornado, this tornado touched down on March 18, 1925, beginning southwest of

The *Chicago Herald Examiner* following the EF5 Tri-State Tornado of 1925.

Airplanes thrown about like toys by the tornado that struck Tinker Air Force Base, Oklahoma, March 20, 1948. COURTESY NOAA, NWS COLLECTION

Ellington, Missouri. The tornado tracked for 219 miles across southern Illinois into southwest Indiana, killing 695 and injuring 2,000 people.

Alfred J. Henry, editor of the *Monthly Weather Review*, described the conditions that occurred with the Tri-State Tornado. Nine of the fifteen rules listed by Finley appeared in Henry's analysis. Finley's work did thus enable forecasters at the time to recognize the possibility for tornadoes, although they were not able to predict the twister in advance.

EXPANDING KNOWLEDGE

Even after the Tri-State Tornado, the word "tornado" continued to be ill-advised, thus deterring tornado research. In 1942 the central areas of the country experienced a large number of tornadoes. Reports indicate that 132 tornadoes killed 223 people and caused just over $7 million in damage. This destruction pointed to the need to warn the public on the possibility for tornadoes.

In the spring of 1943, the Weather Bureau formed experimental tornado warning systems in Wichita, Kansas; Kansas City, Missouri; and St. Louis, Missouri. In June 1944, reporting by local weather observers was expanded to include tornadoes, thunderstorms, hailstorms, lightning, and high winds. Forecasters could make advanced weather forecasts indicating that conditions were favorable for severe storms over a general area, but they could not give, in advance, an exact time or place where severe weather might occur.

FAWBUSH AND MILLER

Research into tornado forecasting continued, building greater understanding into tornado formation. A major breakthrough occurred in the late 1940s, when Major Ernest J. Fawbush and Captain Robert C. Miller of the U.S. Air Force worked on observational and experimental techniques for predicting severe storms and tornadoes. On March 20, 1948, a tornado struck Tinker Air Force Base in Oklahoma City, destroying thirty-two military aircraft and causing $10 million in damage.

Five days later, Fawbush and Miller recognized a similar weather pattern establishing itself over the southern half of the Plains states. Based on their observations, the first tornado forecast was issued and a Tornado Safety Plan for Tinker Air Force Base was put into effect. A tornado ended up striking the base again just after 6:00 P.M. This time the tornado caused $6 million in damage, but nobody was killed and the forecast and the safety plan were considered successful.

While the successes of Fawbush and Miller brought attention from the public and local media, forecasts were only available to Air Force weather offices. New ideas like those presented by Fawbush and Miller were met with skepticism, being challenged and experimented on by other forecasters in the Air Force and Weather Bureau.

THE TORNADO PROJECT

Developments and advancement in the application of radar were made during World War II to detect enemy warships and aircraft. Sometimes the radar images would be obscured by rain, which introduced the possibility of using radar in weather forecasting.

In 1950 the Tornado Project was conducted across Kansas and Oklahoma and included an observational network of 134 stations and 34 cooperative stations. Radar coverage for the project was provided by the Weather Bureau in Wichita, Kansas, and Norfolk, Nebraska, and by the Air Weather Service at Offutt Air Force Base near Omaha, Nebraska; Sherman Air Force Base near Fort Leavenworth, Kansas; and Vance Air Force Base in Enid, Oklahoma.

Sound analysis of data from radiosondes and rawinsondes that started with the Tornado Project became a principal forecasting tool.

CONTINUED RESEARCH

Tornado outbreaks in Missouri, Arkansas, and Tennessee on March 21, 1952, killed more than 150 people and led to public outcry and congressional pressure for the issuance of tornado forecasts. As a result, the Severe Local Storms (SELS) Center was established within the Weather Bureau in June 1953. SELS developed a checklist for forecasting tornadoes that included the basic parameters founded by Fawbush and Miller. This checklist, known as the Lifted Index, became the basis for evaluating potential instability for many decades to come.

Continued research shed new light on the conditions present when tornadoes form. The results of various individual research projects were combined in 1955 to produce a manual for SELS Forecasting Procedures, the first known publication in forecasting severe thunderstorms. This manual would be expanded and its contents debated and argued to produce the first publicly issued document: *Forecasting Guide No. 1: Forecasting Tornadoes and Severe Thunderstorms.*

THE PALM SUNDAY OUTBREAK

The famous Palm Sunday Outbreak occurred on April 11–12, 1965, beginning over parts of Iowa, Minnesota, and southern Wisconsin, and spreading through northern Illinois and Indiana into northern Ohio. At the time of the outbreak, the Weather Bureau had WSR-57 radars and the TIROS VIII satellite images available for forecasting.

Despite these technologies, the Palm Sunday Outbreak killed 271 people and caused $200 million in damage. While SELS was issuing

The 1965 Palm Sunday tornado outbreak was a turning point for the Weather Bureau. A massive double-funnel tornado near Dunlap, Indiana, between Goshen and Elkhart, tore up everything in its path.
COURTESY NOAA, NWS COLLECTION

tornado watches for areas where tornadoes could occur, it was difficult for Weather Bureau offices to communicate with each other and to deliver timely information to media outlets to warn of approaching storms. Checklists developed by Fawbush and Miller, techniques to evaluate air masses, and advances in research and theories were all helpful in understanding where severe weather and tornadoes could occur, but jammed telephone lines made it difficult to deliver advanced warnings and information to the public.

Work continued at the Techniques Development Unit of the National Severe Storms Forecast Center in Kansas City, Missouri, as well as the National Severe Storms Laboratory in Norman, Oklahoma. Forecasts were gradually improving with studies of the Palm Sunday Outbreak and discussions on concepts of storm structure.

In 1972 the U.S. Air Force published a series of guidelines known as "Miller's Rules." Written by Captain Robert Miller, this publication became the main reference for severe weather forecasting in all corners of meteorology, laying down guidelines for weather analysis as well as the use of different symbology for marking severe storm and tornado conditions. The publication also explained forecasting parameters for tornadoes, large hail, and convective wind gusts.

Future work and case studies stemmed from this document. Miller's Rules propelled severe weather and tornado forecasting forward, allowing the expertise shared by Fawbush and Miller to be utilized and expanded by aspiring severe weather forecasters and enthusiasts.

THE F-SCALE

Around the time that Miller's Rules were published, Dr. T. Theodore Fujita introduced the F-Scale, which uses the damage caused by a tornado to estimate its wind speed. Fujita's scale included six levels of tornado intensity, from F0 to F5, and connected tornado damage with the wind scale of the Beaufort scale, developed by Admiral Beaufort of the British Navy in the early nineteenth century.

The Super Outbreak on April 3–4, 1974, set present-day records with 148 tornadoes occurring in thirteen states in a matter of sixteen hours. The tornadoes from this outbreak caused 315 deaths and nearly 6,000

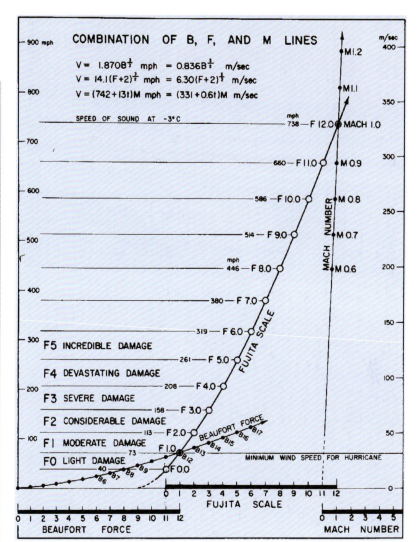

Fujita's logarithmic chart for classifying tornadoes.
COURTESY NOAA, STORM PREDICTION CENTER

injuries. Damage was estimated at $600 million. Fujita and his staff surveyed the damage by land and air, rating every tornado path found using the F-Scale.

When NOAA published *Natural Disaster Survey Report 74-1, The Widespread Tornado Outbreak of April 3-4, 1974*, the F-Scale became the foundation on which all past and future tornadoes would be evaluated. The Techniques Development Unit of the National Severe

Storms Forecast Center hired students to read publications and newspaper articles on tornadoes since 1950 and to assign F-scale damage ratings to these twisters.

Today, tornado damage is assigned a rating based on the worst damage along the tornado's path. However, the original Fujita Scale was

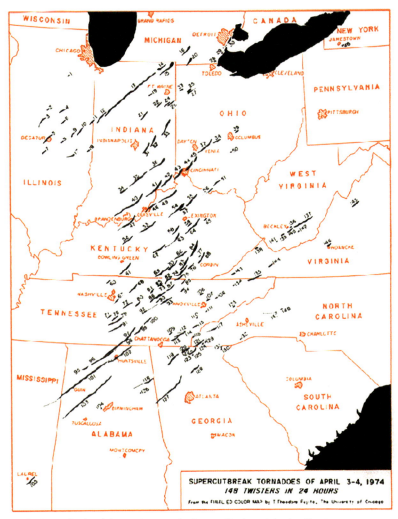
Fujita's analysis of the 148-tornado Super Outbreak in April 1974.
COURTESY NOAA

Cone tornado kicking up debris. COURTESY NOAA, NWS COLLECTION

replaced in February 2007 with the refined and improved Enhanced Fujita Scale (EF-Scale), which incorporates more damage indicators and degrees of damage than the original Fujita Scale, allowing more detailed analysis and better correlation between damage and wind speed. The EF-Scale continues to rate tornadoes on a scale from zero to five, but ranges in wind speed on the scale are more accurate. The EF-Scale still estimates wind speeds, but more precisely takes into account the materials affected and the construction of the structures damaged by the tornado. This rating becomes significant as many of today's forecast techniques are compared to the potential of tornadoes producing EF2 damage or worse.

TORNADO FORECASTING TODAY

A tornado is a violently rotating column of air extending from a thunderstorm to the ground. The most violent tornadoes are capable of tremendous destruction with wind speeds of 250 miles per hour or more. Damage paths can be in excess of one mile wide and fifty miles long.

Continued research and advancements in computer technology from the 1960s through the 1990s improved severe weather and tornado forecasting. Meteorologists were soon able to develop numerical weather prediction models and technology. Projects at organizations like the National Severe Storms Laboratory and National Center for Atmospheric Research in Boulder, Colorado, would assist forecasters in analyzing conditions favorable for severe storms as well as training forecasters to recognize signatures on radar and satellite for improved warnings.

The development, training, and deployment of Doppler radar from the research world into the operational areas of meteorology proved to be the next boost in severe storm and tornado forecasting. Doppler radar enabled meteorologists to not only detect areas of precipitation, but also to detect wind circulations that may develop prior to a storm producing a tornado. Doppler radar enabled the National Weather Service to modernize its operation to provide as complete coverage as possible and serve the public with improved warnings. Today, the National Weather Service has improved warning lead time to an average of fifteen minutes before a tornado is reported.

Today, severe weather forecasters use a combination of Doppler radar, enhanced satellite imagery, and sophisticated analysis programs to rapidly make essential life-saving decisions. Forecasters and researchers have at their disposal, in fractions of a second, a varying array of data that were hand-plotted just forty to fifty years ago. During the Super Outbreak of 1974, average warning lead time was just a few minutes. During the central Oklahoma tornado of May 3, 1999, tornado warnings were available up to a half-hour in advance, and with advancements in broadcast technology, a play-by-play description of developing severe weather was available for hours in advance. Compare the nearly 700 people killed in the Tri-State Tornado in 1925 to the tornado of May 3, 1999, when only forty-four people were killed even though the tornado destroyed nearly 9,000 structures and caused $1 billion in damage.

The first FF5 tornado in the United States since 1999 destroyed the town of Greensburg in southwest Kansas on May 4, 2007. With a thirty-nine-minute warning lead time and extra "tornado emergency" messages issued by National Weather Service forecasters ten to twelve minutes before the storm hit, most residents were able to seek shelter and only eleven people died. Advanced warnings for the Parkersburg EF5 one year later gave residents there a similar advantage.

The hard work, dedication, and knowledge of many scientists all played key roles in advancing severe storm and tornado prediction through seventy years of research and observation. On the horizon is the development of the Phased Array Radar. This new technology will allow researchers and forecasters to analyze storms with much faster electronic scans, leading to improved knowledge of thunderstorm and tornado development and ultimately, even better warnings in the future. ✺

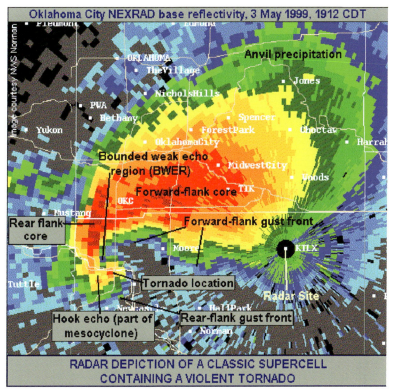

A 1999 Doppler image of an EF5 tornado approaching Moore, Oklahoma. Analyzed by mobile radar, record-breaking winds of 318 mph were measured. COURTESY NOAA

CREDITS LEFT TO RIGHT: SHIRLEY A. MILLER, TOM WAGNER, NATIONAL WEATHER SERVICE, NATIONAL WEATHER SERVICE, DAN BLEVINS

A RARE OCCURRENCE

EF5 TORNADOES AND IOWA TORNADO CLIMATOLOGY (1980–2007)

To achieve a greater appreciation of the strength of the Parkersburg tornado and its relationship to all Iowa twisters, portions of a comprehensive report on Iowa tornado climatology are displayed in Figure A. Prepared by Craig Cogil of the Des Moines National Weather Service, the report clearly shows the rare nature of violent Iowa twisters, with 96 percent of Iowa's 1,302 tornadoes on the low end of the EF-Scale (meaning a five-second gust that blows between 65 and 135 miles per hour). The remaining 4 percent of Iowa's storms were classified EF3 and EF4, with no EF5s in the twenty-eight-year study.

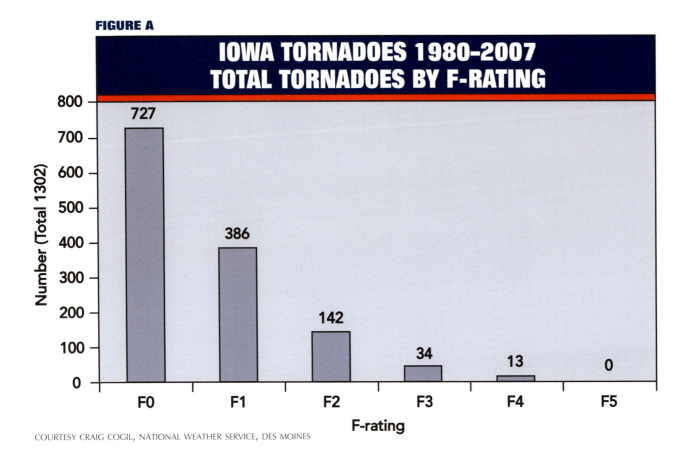

FIGURE A

IOWA TORNADOES 1980–2007
TOTAL TORNADOES BY F-RATING

COURTESY CRAIG COGIL, NATIONAL WEATHER SERVICE, DES MOINES

Parkersburg EF5 struck at 4:59 P.M.

As available energy increases with daytime heating, so too does the number of reported tornadoes, with nearly 60 percent of Iowa's tornadoes occurring between 4 and 8 P.M.

A little-known fact about Iowa is that it is one of the nation's most fertile breeding grounds for EF5 tornadoes. With six touchdowns since the advent of modern records in 1950, Iowa is currently tied with Kansas for the national lead in EF5 production (Figure F). On the same chart, if you refer to the plots of EF5 tornadoes, there is a high incidence of these violent storms in a corridor that extends from south-central Wisconsin into northeast Iowa, south-central Kansas, and north-central Texas. Since 1950 nearly 50 percent of the EF5 storms in the United States have occurred in this area, establishing it as the modern-day "tornado alley" for EF5s.

Cogil's research reveals that the Parkersburg EF5 occurred during statistical peaks for Iowa tornadoes in terms of the time of year (Figure B) and, in particular, the time of day when violent tornadoes are most likely to occur. Twisters need ample energy to form and maintain their structure, which results in a notable spike in tornado activity during peak heating from midafternoon until sunset (Figure C). There is a corresponding increase in violent tornadoes when peak heating is maximized around 5 P.M. (Figure D). Note the significant spike in the number of violent tornadoes, F3 or greater, at 5:00 P.M. This correlates well with statistical peaks in heating and moisture convergence, two key parameters in determining tornado intensity. Interestingly, the

Even more fascinating is the likelihood of EF5 tornadoes developing around Parkersburg and Butler County, Iowa. Since 1966 all five of Iowa's EF5 tornadoes have occurred within fifty-five miles of Parkersburg. Over that period, thirty-two of the state's sixty-seven fatalities have occurred within that radius. For whatever reason, the city of Parkersburg seems to be the epicenter of EF5 storms (Figure E).

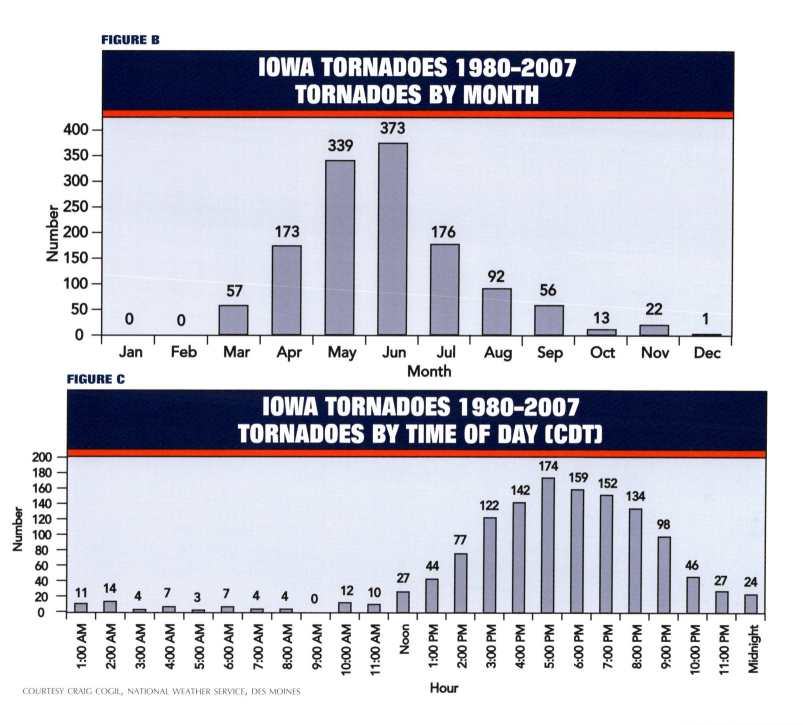
COURTESY CRAIG COGIL, NATIONAL WEATHER SERVICE, DES MOINES

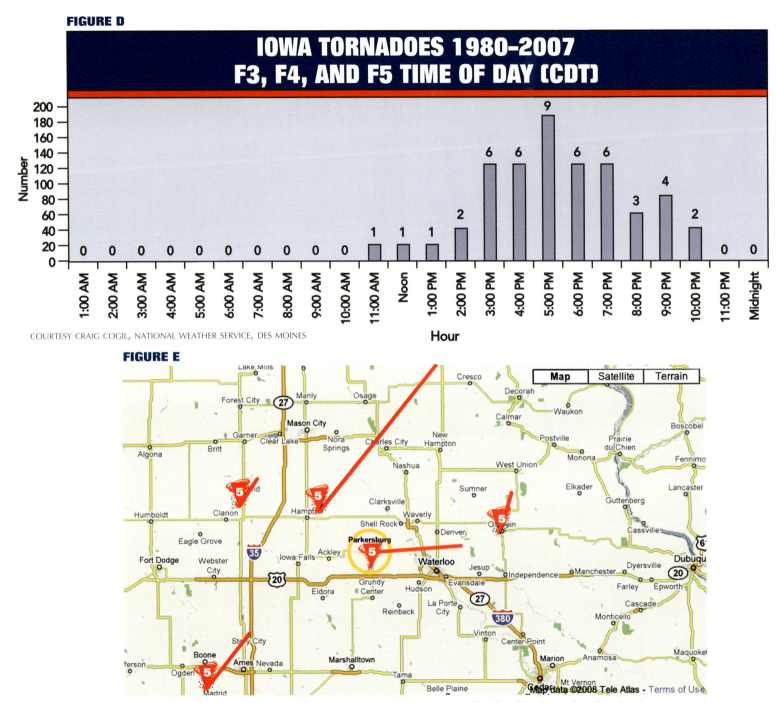

Coincidence or not? Five of the six modern-day tornadoes (1950 to 2008) have occurred within fifty-five miles of Parkersburg.

FIGURE F

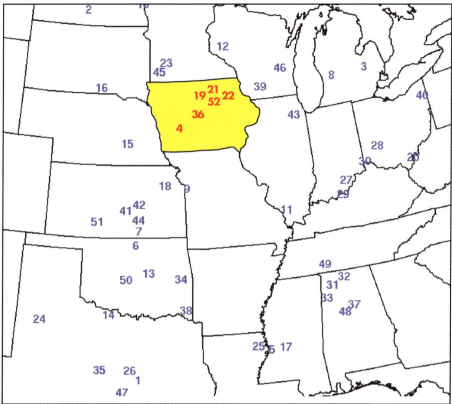

F5/EF5 tornadoes since 1950. COURTESY NWS STORM PREDICTION CENTER

TORNADO SAFETY

PREPARED BY NOAA AND THE NATIONAL WEATHER SERVICE

What you need to know NOW! Listen to NOAA Weather Radio All Hazards or commercial radio/television for tornado warnings and instructions. If you hear a tornado warning seek safety immediately.

INDOORS:
* Abandon mobile homes—they are not safe even when tied down. Go to a designated shelter.
* Go to a basement or interior room on the lowest floor (bathroom or closet without windows, under stairs). Get under a sturdy piece of furniture.
* Cover yourself with a mattress or blanket.
* Put bicycle helmets on kids.
* Put on sturdy shoes.
* Put infants in car seats (indoors!).
* If you have time, gather prescription medications, wallet, and keys.
* DO NOT open your windows!

IN A VEHICLE:
* Leave the vehicle for sturdy shelter or drive out of the tornado's path.
* DO NOT hide under overpasses—they provide no shelter.
* Lie flat in a ditch (last-resort).

OUTDOORS:
* Find a culvert or cave.
* Find something to hang onto.
* Lie flat in a ditch.
* Cover your head.

By virtue of its land area and location, Oklahoma City has been hit by more tornadoes than any other city. The worst of these was an EF5 that struck on May 3, 1999, causing thirty-six deaths and a billion dollars of damage.

From 1980 to 2007, Iowa experienced 1,302 confirmed tornadoes. The most reported in a single year was 120 in 2004. The least was 22 in 1996.

LIKE A ROCK

Parkersburg, a town of 1,900 people, has eight churches. This farming community is built on a rock of faith . . . a boulder so large not even a killer tornado could budge it. Eight miles away sits New Hartford, a community of 600, similar in faith and hope; it, too, came face to face with the storm. The strength of both towns' survivors is seen everywhere. From a shredded American flag stuck jauntily in a pile of rubble to a man standing in the wasteland of his home saying, "I still have my wife. I have my children. I am a lucky man."

In retrospect, this storm is a story about faith and triumph, of people who can count on their neighbors, who live in a state that shares its resources, in a country where thousands of volunteers poured in from every corner. So great was the response from outside assistance that 1,240 volunteers registered to help in a single day. They came with skid loaders, chainsaws, and buckets. They got down on their hands and knees and picked up garbage.

Two days after the deadly storm passed, Pastor Donna Ruggles stood before her congregation at the Bethel Lutheran church and delivered what was perhaps her most emotional message. Her parishioners had suffered heavy losses, with forty-four family homes destroyed. She spoke of the tears she had witnessed in the eyes of an older member because physically he could no longer help with the clean-up. She talked of a father and son insurance agency who were attempting to aid the entire community, and how the weight had become almost too much to bear. She spoke of others who had not been affected, giving of themselves to the breaking point. Finally, she shared how she had held many in her arms, standing in the rubble of what had once been their homes, and of how they had prayed together.

She also noted that anger was a natural outcropping of such a disaster. She then directed an answer to those who questioned where God was that Sunday by saying, "He was here as you were dragged from your home by someone, perhaps even a stranger. He was here when you were taken into the home of someone you did not know, so that you had a place to stay that night. He was here when your community leaders worked around the clock to get the town organized out of chaos, and he will be with you today and in all the days ahead in the decisions that have yet to be made."

The stories of heroism, courage, and selflessness are echoed over and over again around Parkersburg and New Hartford. It is inspirational to hear people who have lost so much find positive endings in every tale. "Did you know . . ." they will ask, and then talk about the new siren system that had gone on line just days before the disaster, the holiday weekend that kept many families out of town, and the empty school buildings because classes were not in session that day. Looking at the devastation, it is clear that loss of life could have been much worse.

This prevailing sense of optimism is seen in the number of residents who are choosing to rebuild their homes and lives in Parkersburg and New Hartford. Of the approximately 300 houses destroyed, 90 percent of those families are staying. It is simple, many say. This is their home, their past, and this will be their future.

New Hartford and Parkersburg were established over 150 years ago when settlers put their faith in a land of fertile fields and rolling hills. Today, while the foundations of both communities may have been shaken, their rock of faith still stands strong.

Parkersburg's angel of mercy, bruised but not broken. COURTESY TIM WILCOX

HOPE AND HEROES AMONG THE HURT
BY U.S. SENATOR CHUCK GRASSLEY

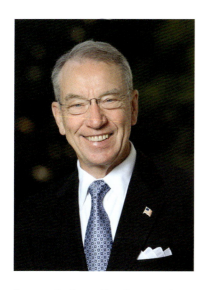

On the evening of May 25, 2008, Mother Nature unleashed a tragic beginning to summer. We didn't know it that night, but it was only the beginning of a month-long string of the kinds of natural disasters that make people realize the perils of pettiness and appreciate what really matters most.

That Memorial Day weekend, a history-making twister produced winds in excess of 200 miles per hour. The EF5 tornado tore across Butler, Black Hawk, Delaware, and Buchanan counties. Then, only a few weeks later, floodwaters that exceeded 500-year flood levels covered much of eastern Iowa and left tens of thousands of Iowans scrambling for food and shelter. And, who could forget the heroic efforts of the Boy Scouts in western Iowa who helped their fallen members when a killer tornado ripped through the Little Sioux Scout Ranch.

Natural disasters have wrought havoc on humanity since the beginning of time. In recent years, the 2004 tsunami in Southeast Asia claimed more than 100,000 lives and displaced millions of victims from their homes. In September 2005, a Category 5 hurricane ravaged America's Gulf Coast, causing $11.3 billion in damage. Last year in Greensburg, Kansas, a tornado leveled the entire community of 1,400, causing an estimated $267 million in damage.

The financial estimate of damage from the tornadoes, storms, and flooding hasn't been calculated. But the price tag won't do justice to the heartbreak and hurt.

Whether it's an earthquake, hurricane, tornado, or flood, a natural disaster leaves behind massive debris and destruction. The physical and financial tolls shouldered by the victims arguably pale compared to the emotional scars and personal losses left in the aftermath of a killer natural disaster.

From my own yard in New Hartford, I watched the dark storm clouds blacken the sky as the tornado made its way across Butler County from Parkersburg to New Hartford. Then, only weeks later I saw 240 of the 270 houses of my hometown under water.

Since that dreadful day when a tornado struck Parkersburg and the following weeks when Iowa was besieged with floodwaters, I've seen nothing more gratifying than the people of these communities pulling together.

The clean-up operation will take months and even years of back-breaking work. But, this seemingly impossible task is being made possible thanks to the tireless commitment of first responders, administrators, emergency crews, and volunteers from across the country.

The outpouring of support from neighbors, friends, and strangers from near and far has given a jump-start to the necessary healing process. It underscores the decency of human nature rising above the catastrophic forces of Mother Nature.

The selfless sacrifice by literally scores of heroes will help mend the immeasurable heartbreak and hurt I saw during my visits to these communities. I say with gladness in my own heart, neither the tornadoes nor the floods extinguished the hope and pride of the residents of these Midwestern communities. And, when we have time to step back and take a deep breath, there will be a new appreciation of just how lucky we are.

Chuck

CREDITS LEFT TO RIGHT: TRAVIS MUELLER, LUANN SWAILS, KATHI JOHANSEN, TRAVIS MUELLER, JENNY HAGARTY

ONCE UPON a TIME

The Parkersburg tornado wreaked its havoc in a matter of seconds, but the disaster for the victims of the Flood of 2008 seemed endless, as the water steadily rose for days on end. Whatever poison you pick—an EF5 tornado or a 500-year flood—in the end the results are eerily similar. Lives are disrupted, neighborhoods destroyed, and familiar landscapes changed forever. Now that you know the ordeal of Parkersburg and New Hartford, it is time for the epic story of eastern Iowa's battle with the mother of all floods.

The Great Flood of 2008 was so exceptional, so out of the realm of what's ever been witnessed, that it is now in a class by itself. So remarkable was the event that, in places like Cedar Rapids, a similar flood is not likely to occur for generations, perhaps even centuries. In fact, during the height of the flood, the volume of water flowing through the city far exceeded the expectations of a 500-year flood. If hydrologists made projections for a 600- or 700-year flood (which they don't), an argument could be made that, in places, the Great Flood of 2008 reached or even exceeded those thresholds. An unprecedented group of conditions came together to create this historic event and, in that regard, created eastern Iowa's "perfect flood."

How do you create the perfect flood? It's harder than you might think. For the answer, it's worth going back to the ice age that began about 2.5 million years ago and concluded nearly 14,000 years ago. During that period, massive ice sheets from the north advanced and retreated successively across Iowa. At their peak, these enormous sheets were one to two miles thick and covered hundreds of thousands of square miles. When they melted, they released tremendous amounts of water that formed huge glacial lakes and rivers. Eventually the lakes drained; the torrents of escaping water formed the valleys through which Iowa's rivers run today.

Over time, the river valleys and fertile soils deposited by the glaciers brought EuroAmerican settlers to Iowa, primarily for fur trading and farming. As settlements pushed across the state, agriculture rose to dominance. Following the Civil War, the expansion of railroads enabled farm produce to reach Eastern markets, bringing about a shift from growing wheat as a cash crop to growing corn to fatten livestock. Food processing developed as the primary manufacturing activity. Soon, cities such as Iowa City, Cedar Rapids, and Waterloo were thriving along the banks of the state's major rivers.

While the rivers provided power, transportation, and drinking water, they occasionally overflowed from melting snows and too much rain. It soon became apparent for those living along the river that seasonal floods were an ongoing threat, and reports began showing up in Iowa's newspapers. This 1851 account is an excerpt from a 1934 edition of the *Palimpsest*: "The deluge began in May. For more than forty days the rain fell, not continuously but at very frequent intervals. Farmers in the valleys despaired of getting their corn planted; Crops of small grain were washed out or ruined. Not until July did the skies clear

The Iowa, Cedar, and Wapsipinicon rivers.

Glacial advances that shaped Iowa's river valleys.
COURTESY IOWA DEPARTMENT OF NATURAL RESOURCES

Eastern Iowa river basins. COURTESY U.S. ARMY CORPS OF ENGINEERS

and the floods subside. A newspaper reported that neither the 'memory of the oldest inhabitant' nor 'any traditional accounts from the Indians' furnished any evidence of such an inundation.

"There was no need of so much rain in Iowa. During the previous year of 1850 the rainfall was estimated at forty-nine inches which, according to modern records, was about eighteen inches above normal. The ground-water level must have been high in the spring of 1851. After the first downpour, the earth became saturated and the surplus ran off to swell the creeks and send the rivers surging out of their banks, even above the second terraces.

"Everywhere the same conditions prevailed, even on the narrow watershed of the Missouri slope. But the damage was greatest in southeastern Iowa, for there the water in the Cedar, Iowa, Skunk, and Des Moines rivers, drained from two-thirds of the State, reached the highest mark. Moreover, that region was the most densely populated portion of Iowa. The settlers, clinging to the valleys, had not penetrated to the upland prairies of the central and northern sections. When the floods came, they discovered their lowland farms were unfavorably situated. Most of the towns, being located on the rivers, were under water, but the inland communities were comparatively unharmed."

That flood still measures as one of the worst on record in Iowa. In Cedar Rapids, the 1851 crest is the second highest on record; in Iowa City, 1851 weighs in at number three.

Despite such catastrophic events, man's undaunted desire to live and build near rivers continued into the twentieth century. In 1918, another significant flood swamped eastern Iowa. In Iowa City, with no reservoir to hold back the Iowa River, it poured through town on June 6 in response to an exceptionally rainy spring. Iowa City historian Irving Weber

Riverside Park, Cedar Rapids, circa 1900.
COURTESY CARL & MARY KOEHLER HISTORY CENTER

Burlington Street bridge, 1918. COURTESY CEDAR FALLS HISTORICAL SOCIETY

3rd Ave and 1st Street W, Cedar Rapids, 1929.
COURTESY CARL & MARY KOEHLER HISTORY CENTER

12th Ave and 3rd Street S.E., Cedar Rapids, 1929.
COURTESY CARL & MARY KOEHLER HISTORY CENTER

Mays Island, Cedar Rapids, 1929.
COURTESY CARL & MARY KOEHLER HISTORY CENTER

3rd Ave and 1st Street W, Cedar Rapids, 1929.
COURTESY CARL & MARY KOEHLER HISTORY CENTER

1944 Cedar Falls. COURTESY CEDAR FALLS HISTORICAL SOCIETY

1961 Highway 218 Cedar Falls. COURTESY CEDAR FALLS HISTORICAL SOCIETY

remembers the flood thus: "During the 1918 flood the lower area of City Park was twelve to fifteen feet under water. At its crest, water flowed over the wood planking of the iron-arched Park Bridge. Probably the greatest disaster of all was the fate of the three large Englert Ice Houses, which upended—loaded with 9,000 tons of Iowa River ice. It was the first time in history the Iowa River was filled with ice in June."

Iowa City was without power for three days.

In March 1929, a late winter flood swept through eastern Iowa, with significant impacts on the Cedar River from its headwaters downstream into Waterloo, Vinton, and Cedar Rapids. The Evening Edition of the *Cedar Rapids Gazette* gave this description of the passing crest, which until 2008 was tied with 1851 as that city's all-time highest: "Cedar

Waterloo Way Wins

Waterloo Daily Courier
FIRST WITH THE NEWS

THURSDAY'S WEATHER

3/29/61

ESTABLISHED 1858 — WATERLOO, IOWA, WEDNESDAY, MARCH 29, 1961 — THIRTY-TWO PAGES — PRICE SEVEN CENTS

Record Flood Slowly Begins Receding At Waterloo; May Be Disaster Area
Bridges and Dikes Still Big Concern to Officials

especially on the east side, were warned to prepare for water in their basements. During the middle of the crest and through much of the day, Virden Creek overflow kept volunteers busy and caused some

COURTESY *WATERLOO DAILY COURIER*

Newly completed Coralville Reservoir in 1957.
COURTESY U.S. ARMY CORPS OF ENGINEERS

Filled and functioning—looking north towards the reservoir and Coralville Lake.
COURTESY U.S. ARMY CORPS OF ENGINEERS

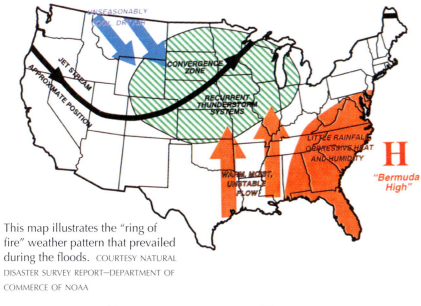

This map illustrates the "ring of fire" weather pattern that prevailed during the floods. COURTESY NATURAL DISASTER SURVEY REPORT—DEPARTMENT OF COMMERCE OF NOAA

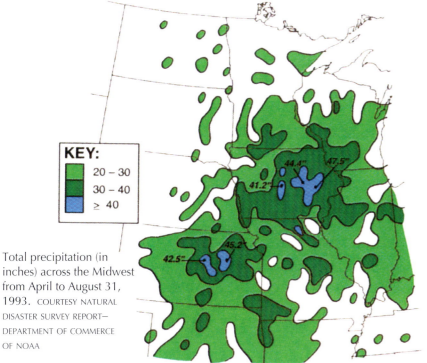

Total precipitation (in inches) across the Midwest from April to August 31, 1993. COURTESY NATURAL DISASTER SURVEY REPORT—DEPARTMENT OF COMMERCE OF NOAA

Rapids, relaxed from 48 hours of anxiety, began today to take stock of its loss caused by the most disastrous flood within the memory of the oldest inhabitant. The relaxing process began around midnight last night when the restless waters began slowly to recede. At 1:30 P.M. today at the gauge at the Hubbard Ice company above the damn showed a drop of 6 inches, and at the Iowa Windmill and Pump company, below the damn, 8 inches. The river was falling at about an inch per hour. At the same time the river had dropped more than 3 feet at Vinton.

"With falling waters came better opportunity to estimate what damage the mud and filth in the swirling torrents had done to merchandise, equipment, and basements. Conservative estimates placed the loss at several hundred thousand dollars."

For the next thirty years, most of eastern Iowa's major rivers had their ups and downs without causing much in the way of significant disruptions. However, as the socially turbulent 1960s arrived, so too did flooding, especially along the Cedar River and its tributaries. The flood of 1961 produced an all-time record crest in Waterloo, and downstream the river's high mark in Cedar Rapids fell less than a half-foot short of the all-time crest.

Many homes in the vicinity of Waterloo and Cedar Falls were evacuated, and up to 1,000 people were reported homeless by the *Waterloo Daily Courier*. The March 29 edition of the *Courier* had this to say about the high water: "The worst flood in Waterloo history was receding Wednesday afternoon, leaving the city in a state of emergency with weakened and damaged bridges, hundreds of acres flooded and perhaps thousands homeless." Iowa's Governor Norman Erbe said he would "investigate the possibility of having the Waterloo and Evansdale areas declared a disaster area so that federal funds might be available for rehabilitation."

April 1965 brought another round of serious flooding to the banks of the Cedar River. Unlike many earlier floods, this event was driven by rapid snowmelt, rather than prolonged heavy rains. That March had been one of the coldest and snowiest on record for many Midwest locations. At month's end, ice covered most lakes and rivers from Iowa north. In early April, the spring thaw arrived with a vengeance, and melting snow and ice sent a wall of water down eastern Iowa's rivers. Crests in both Waterloo and Cedar Rapids were just short of those experienced in 1961. The flood was of such magnitude in Waterloo that the water came within two inches of breaking 1961's all-time record. Cedar Rapids measured its fifth-highest crest on April 10.

Following the floods of the 1960s, congressional flood-control acts enabled the U.S. Army Corps of Engineers to design and construct water development projects to help control flooding on many of Iowa's rivers. Along the Iowa River, a major project with similar goals had already been completed on September 17, 1958, with the grand opening of the Coralville Lake reservoir. The dam and its 712-foot-high spillway were constructed to create a vast recreational lake and to control flooding at downstream locations such as Iowa City and Coralville.

Over the next eighteen years, eastern Iowa's rivers were fairly well behaved. The years 1969, 1974, and 1991 all saw some degree of flooding, but the lion's share of problems were confined to the Iowa River. Even there, the floods were confined to primarily agricultural locations upstream from the Coralville Dam.

By the end of 1992, a unique and extreme set of meteorological conditions came together to produce what would become the Great Flood of 1993. The previous wet fall had resulted in above-normal soil moisture and water storage conditions for much of the upper Midwest. These conditions were followed by a nearly stationary weather pattern that kept an active storm track locked in place throughout the spring and summer of 1993. Storm systems with repetitive heavy rains and broad coverage bombarded the upper Midwest. Some parts of east-central Iowa received nearly fifty inches of rain during that period.

For several months, swollen rivers raged throughout Iowa and surrounding states. When the waters receded, the flood of 1993 had surpassed all floods in modern U.S. history in terms of the amount of record crests, overall coverage, persons displaced, and crop and property damage, as well as duration. For the first time in its thirty-five-year history, water poured over the spillway of the Coralville Dam, allowing the Iowa River to reach an all-time crest in Iowa City. In both Waterloo and Cedar Rapids, the Cedar River crested at its fourth-highest level.

Following the flood of 1993, an executive summary issued by NOAA stated, "The duration, extent, and intensity of the flooding defines this event in the twentieth century. Measured in terms of economic and human impacts, the Great Flood of 1993 will be recorded as the most devastating flood in modern U.S. history." It was widely believed by experts and people throughout Iowa that a flood of that scope would not be seen again for decades, perhaps as long as a century.

DEFYING THE ODDS

Just fourteen years later, subtle signs of another historic flood began to emerge. By the time it swamped the soils of eastern Iowa, it would prove

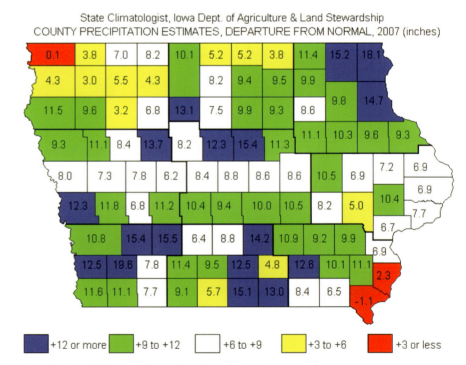

Most of Iowa's 2007 rainfall was 6 to 12 inches above normal.
COURTESY STATE CLIMATOLOGIST HARRY HILLAKER

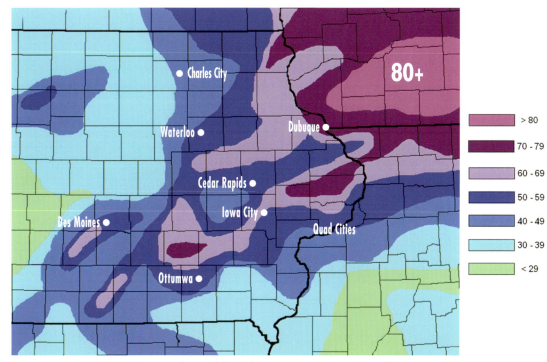

2007–2008 seasonal snowfall totals. COURTESY NWS QUAD CITIES (DAVID SHEETS)

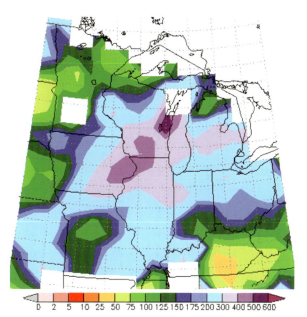

Total snowfall percent of mean, February 1, 2008 to February 29, 2008. COURTESY MIDWEST REGIONAL CLIMATE CENTER

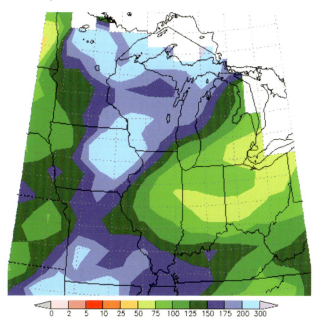

Total precipitation percent of mean, April 1, 2008 to April 30, 2008. COURTESY MIDWEST REGIONAL CLIMATE CENTER

to be even worse than 1993. The foundation of this new flood would be precipitation, and lots of it, in 2007. More than a year before the all-time crests of 2008, rain and snow fell often and in earnest around the state. By the time 2007 came to a close, it ranked as the fourth-wettest among 135 years of Iowa records. The soils of eastern Iowa resembled a leaky sponge that needed to be squeezed.

The winter of 2007–2008 was unusually cold and wet. Snow began piling up early in December and, by March, much of northeast and east-central Iowa had record or near-record snowfall of fifty-five to seventy-five inches. A red flag was flying in regard to the potential for spring flooding. Jeff Zogg, service hydrologist at the National Weather Service in the Quad Cities, was concerned. "As hydrologists, we are very interested in the water equivalent of the snow," he said. "At its peak, we verified five inches of moisture within the snowpack, with potential amounts of up to seven inches. If it melts rapidly, that would be the equivalent of

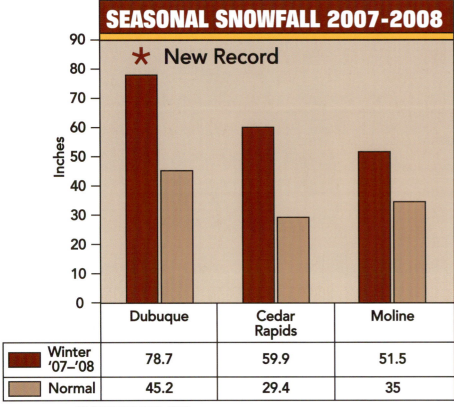

COURTESY NWS QUAD CITIES

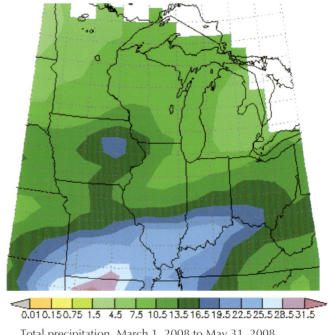

Total precipitation, March 1, 2008 to May 31, 2008.
COURTESY MIDWEST REGIONAL CLIMATE CENTER

dropping a lot of water on eastern Iowa's rivers." With Iowa's winter the eighth-wettest on record and the previous year such a soaker, the stage was set for a serious flood.

Fortunately, the spring thaw arriving in March was essentially ideal for snowmelt. Temperatures rose above freezing during the day and fell below during the night. Precipitation was actually below normal. The snowpack gradually diminished, and with it went the threat of significant flooding. Zogg knew he had dodged a bullet but, he noted, "The melting snow had added moisture to the ground and both soil moisture and river levels were running well above normal."

The break in the foul weather did not last long, and the stormy pattern that prevailed during the winter was soon reestablished. Clouds and rain were regular visitors; the combination kept temperatures uncomfort-

ably cool and well below normal. During April, the mercury reached 70 degrees F just four times in Cedar Rapids. In Waterloo, the first 80-degree day of 2008 was not felt until May 6. Just north of the Iowa border, Rochester, Minnesota, reported a trace of snow on May 3. Overall, the spring was the twenty-ninth coldest and twelfth-wettest in Iowa history. State climatologist Harry Hillaker said the cool conditions contributed to the upcoming floods because "the lower temperatures resulted in lower evaporation rates, thus slowing the rate of drying."

Another key impact of Iowa's late spring was the large reservoir of cold air that still lingered over the upper Midwest. With the coming of summer, warm muggy air was advancing northward and a strong storm track was establishing itself between the two air masses. A classic weather pattern known for its ability to produce violent thunderstorms and torrential rains was in the making. The severe weather season, which thus far had been quiet as a mouse, was getting set to roar.

The straw that broke the back of eastern Iowa's rivers was a two-week stretch of tumultuous weather that commenced with the Parkersburg

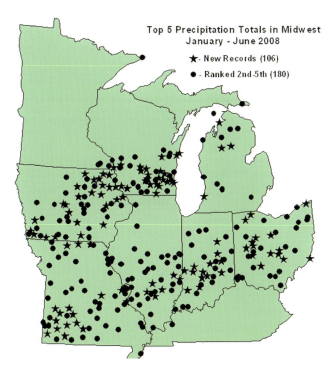

COURTESY MIDWEST REGIONAL CLIMATE CENTER

tornado on May 25. During the last week of May, clusters of heavy rain producing thunderstorms roamed southern Minnesota and northern Iowa. Rainfall amounts up to 400 percent above normal fell over the soggy headwaters of the Cedar, Iowa, and Wapsipinicon rivers. Accordingly, all three rivers began to rise at an alarming pace.

At the National Weather Service, Zogg and his staff prepared for a significant flood, but not necessarily a historic one. However, once the rains began they were not about to stop. From May 25 to June 15, rain fell nineteen out of twenty-two days in Cedar Rapids. With each passing deluge, concerns of flooding continued to mount. Zogg's office began hosting daily flood coordination conference calls on Tuesday, June 3. Attendees to these calls were the U.S. Army Corps of Engineers (COE), FEMA, the States of Iowa and Illinois (including their respective Homeland Security Departments), the NWS North Central RFC (NCRFC) and NWS Des Moines. Zogg indicated that "during each call we summarized the current hydrologic conditions and provided our thoughts and concerns

---

BULLETIN - IMMEDIATE BROADCAST REQUESTED
FLOOD WARNING
NATIONAL WEATHER SERVICE QUAD CITIES IA IL
1146 AM CDT SUN JUN 08 2008

...FORECAST FLOODING CHANGED FROM MINOR TO RECORD SEVERITY FOR THE FOLLOWING RIVERS IN IOWA...
  CEDAR RIVER AT VINTON AFFECTING BENTON AND LINN COUNTIES
  CEDAR RIVER AT CEDAR RAPIDS AFFECTING CEDAR...JOHNSON AND LINN COUNTIES

...FORECAST FLOODING CHANGED FROM MAJOR TO NEAR RECORD SEVERITY FOR THE FOLLOWING RIVERS IN IOWA...
  CEDAR RIVER NEAR CONESVILLE AFFECTING CEDAR...LOUISA AND MUSCATINE COUNTIES
  IOWA RIVER AT MARENGO AFFECTING IOWA AND JOHNSON COUNTIES
  IOWA RIVER AT COLUMBUS JCT AFFECTING LOUISA COUNTY
  IOWA RIVER AT WAPELLO AFFECTING LOUISA COUNTY

...VERY HEAVY RAINFALL OVER THE PAST SEVERAL DAYS WILL LEAD TO SIGNIFICANT RISES...ABOVE FLOOD STAGE...ON THE CEDAR AND IOWA RIVERS IN EASTERN IOWA. RECORD OR NEAR RECORD FLOODING IS EXPECTED IN SOME LOCATIONS. THIS IS A SERIOUS SITUATION. MONITOR SUBSEQUENT RIVER FORECASTS... AS WELL AS FLOOD STATEMENTS AND WARNINGS...FOR THE LATEST INFORMATION.

COURTESY NWS STORM PREDICTION CENTER

regarding river forecasts." Initially the conference calls began with city officials in Mason City and Charles City and were later extended to other cities on the Iowa and Cedar rivers as the crisis escalated there.

On the 7th and 8th of June, a massive thunderstorm complex inundated northern Iowa with three to nine inches of rain. As the water poured into the headwaters of eastern Iowa's rivers, the Flood of 2008 gained new steam, respect, and heightened attention. Flooding forecasts abruptly changed from minor to record severity along the Cedar and Iowa rivers. ✸

CREDITS LEFT TO RIGHT: TRAVIS MUELLER, KWWL SHOW AND TELL 7, TRAVIS MUELLER, TRAVIS MUELLER, DENNY BOWMAN

CEDAR RIVER: the FIRST TO GO

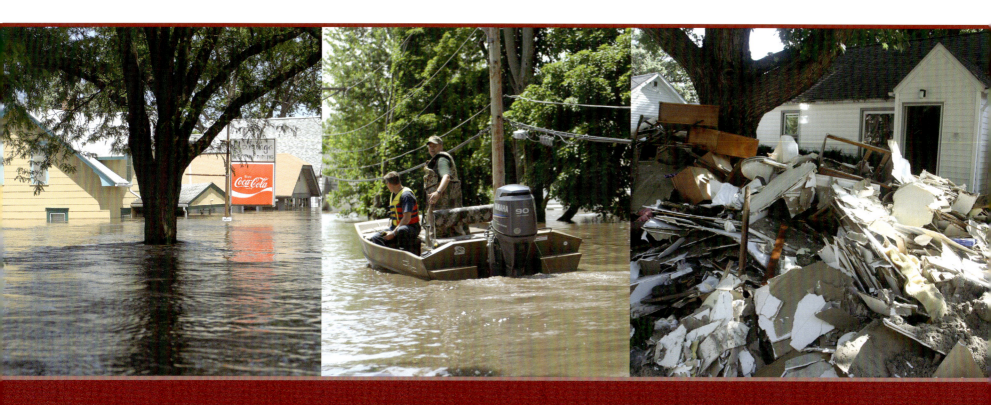

Charles City and Waverly

As the water was absorbed into the soggy river basins of northern Iowa, it did not take long for record crests to be established on the Cedar and Shell Rock rivers between June 8 and 10 from Charles City southward through Waverly, Shell Rock, and Janesville.

With the waters surging south toward Cedar Falls and Waterloo, Zogg found himself in a predicament as the Cedar and Shell Rock rivers and their smaller tributaries merged just north of Cedar Falls. Because each river and tributary had its own crest, it was difficult to get a handle on how they all would interact. Would there be individual crests or one big one? According to Zogg, "They all came together at once and produced a really high crest, it was the worst case scenario." He added, "Things were really going downhill fast."

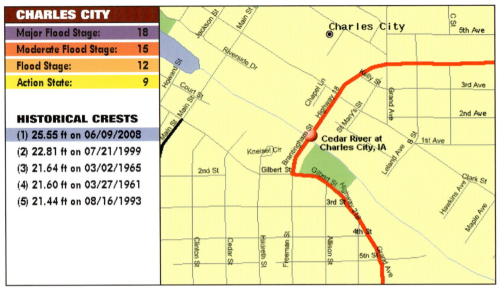

COURTESY NATIONAL WEATHER SERVICE

Downtown Waverly bridge looking east. COURTESY DENNY MILLS/GAYLON ISLEY

In many places, the water was literally rising before people's eyes. Rain was occurring so frequently that models used to make crest estimates were having a hard time keeping up. So instead of running the models twice a day, which is standard, hydrologists increased the number of run times to three per day and eventually upped that figure to six. The crest projections produced were based on recent rainfall, predicted rainfall, and the amount of water already measured at staging points. The crest numbers were climbing higher, and Zogg and his fellow workers reported to emergency managers that "this doesn't look good and that this is going to be a really serious situation." A special flood statement issued by the National Weather Service at 10:22 A.M. on June 9 echoed those sentiments.

On Monday, June 9, when news broke that moderate to severe flooding was expected in the city of Waverly, the response was immediate. "All day long it was hectic," said Mike Cherry, Waverly's city engineer. "It is amazing how quickly a community can rise and work together to secure homes and properties," he added. The city worked at a fever pitch to distribute 200,000 sandbags, with many city employees choosing to help their city rather than protect their own homes. "About one-fourth of our employees were personally affected by the floodwaters," said Waverly city administrator Richard Crayne. "The majority of them made the decision to stay at work and help with the city's sandbagging efforts. So their families were left trying to prepare and secure their properties on their own."

That morning, along with the sandbagging efforts, city leaders ordered a mandatory evacuation for about one-fourth of the community of 9,400 residents. The Waverly Fire Department canvassed door-to-door, alerting residents to evacuate their homes by eight that

FLOOD STATEMENT
NATIONAL WEATHER SERVICE QUAD CITIES IA IL
1022 AM CDT MON JUN 09 2008

...RECORD FLOODING IS EXPECTED...THIS IS A SERIOUS SITUATION WITH UNPRECEDENTED RIVER LEVELS LIKELY...
HEAVY RAINFALL WILL LEAD TO **MAJOR TO RECORD FLOODING** ALONG PORTIONS OF THE CEDAR AND IOWA RIVERS IN EASTERN IOWA. MINOR FLOODING IS EXPECTED ON THE ENGLISH RIVER. PEOPLE ALONG PORTIONS OF THE CEDAR AND IOWA RIVERS SHOULD PREPARE FOR UNPRECEDENTED RIVER LEVELS IN SOME AREAS.

PLEASE NOTE...RIVER FORECASTS FOR THE IOWA RIVER WILL BE UPDATED LATER TODAY AS THE U.S. ARMY CORPS OF ENGINEERS PROVIDES THE LATEST PLANNED OUTFLOW INFORMATION FOR THE CORALVILLE RESERVOIR.

COURTESY NWS STORM PREDICTION CENTER

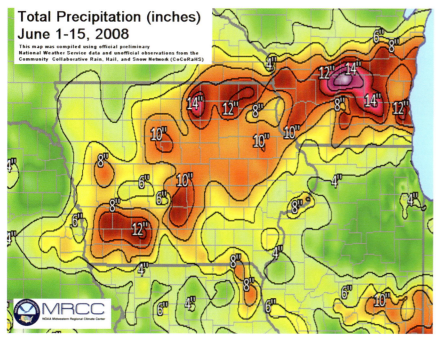

COURTESY MIDWESTERN REGIONAL CLIMATE CENTER

night. At that time, gas and electrical service to the affected neighborhoods would be shut off.

By nine o'clock that Monday evening, the National Weather Service revised the crest projection from 16.8 feet to 19 feet. It was the worst possible news. The crest would be two feet higher than Waverly's all-time record set in 1999. "We went from a 100-year flood event to a 500-year flood event," said Cherry.

Early in the morning on June 10, the phones started ringing at the emergency evacuation center. "Two A.M. was like the witching hour," said Cherry. "People who did not heed the evacuation order started to call. They wanted the city to come and evacuate them." With floodwaters reaching the first floor of many homes, it was deemed too risky to attempt night rescues. Residents were told to wait for daylight. "The danger of doing water rescues in the dark is that you can't see anything. The water can be over the top of fire hydrants and there is other debris in the water that you are not aware of," Cherry explained.

When daylight broke Tuesday morning, floodwaters were rushing through town, spilling into City Hall and the Waverly Civic Center, closing three of the city's bridges. By Tuesday afternoon, the river had crested at its new record of 19.1 feet. In all, 700 to 800 homes were impacted, and up to fifty businesses. In retrospect, Crayne said the sandbagging was not time well spent. "Even if you were able to keep the water out of your property it would not have helped, because the ground pressure would more than likely have caused foundation failures."

As the water began to recede, tensions continued to escalate. On Wednesday night, June 11, rain fell to the north of the community and the National Weather Service issued a warning for a second projected crest of sixteen feet for Saturday evening. "Everyone started holding their breaths," said Crayne. "Our concern was with people putting debris on the streets and having it flood again . . . all that extra debris floating around could possibly cause injury or damage."

Townsfolk continued to worry as the river's level fluctuated through the next few days, but finally, on Sunday morning, came a welcome relief. The Cedar had crested for the second time and, with that crest, only minimal flooding was reported.

Charles City. COURTESY CHERA REGAN

Waverly. COURTESY KWWL SHOW AND TELL 7

Waverly. COURTESY MARK SWINTON

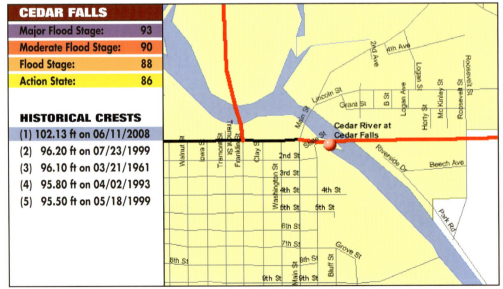

CEDAR FALLS	
Major Flood Stage:	93
Moderate Flood Stage:	90
Flood Stage:	88
Action State:	86

HISTORICAL CRESTS
(1) 102.13 ft on 06/11/2008
(2) 96.20 ft on 07/23/1999
(3) 96.10 ft on 03/21/1961
(4) 95.80 ft on 04/02/1993
(5) 95.50 ft on 05/18/1999

COURTESY NATIONAL WEATHER SERVICE

WINNING THE BATTLE
Cedar Falls

In the first week of June, Cedar Falls city officials learned the Cedar River was expected to crest between ninety-four and ninety-eight feet, possibly putting it above the 1999 record of 96.2 feet.

By Monday, June 9, the predicted crest rose to 100 feet. Despite that, city leaders remained confident the levee system, built to an elevation of 103 feet, could withstand the rising tide.

The morning of June 10 dawned clear and bright. The city was experiencing the first beautiful day seen all spring. But with the good weather came bad news. "I was told the new flood forecast had come out at one hundred and three. I said, 'You mean one hundred point three?' and was told 'No, one hundred and three feet!'," said Director of Development Services Ron Gaines. "We decided to get everything in place before we issued an announcement. The sun was shining, and if people heard what was

Cedar Falls. COURTESY DENNY MILLS/GAYLON ISLEY

Manning the line in Cedar Falls. COURTESY BILL WITT

Cedar Falls. COURTESY BILL WITT

happening, they would want to come down and see," he added, referring to the sightseers that floods typically bring.

City leaders scrambled to prepare for a mandatory evacuation of the downtown. Contractors were called and told to bring in sand and equipment to help support the levee system. Bus transportation was arranged to bring sandbaggers in and out of the affected area.

By afternoon, most of the bridges on the Cedar River north of Cedar Falls and Waterloo were either closed or washed out; a trip across the river that should have taken fifteen minutes was now a nightmare of detours that added hours to a journey.

Meanwhile, 3,000 to 4,000 volunteers, including young children, college students, and retired citizens, came out to bolster the levee with sandbags. In just twelve hours, a wall of sandbags three feet high was placed on top of the existing mile-long levee. In addition, city engineers set up twenty-four-hour monitoring of the levee, walking up and down, checking for weak spots or trouble areas. The biggest concern was that the levee would become so saturated that water would burrow underground, create a tunnel, and explode on the other side—forcing the levee to breach.

A call from Craig Witry, a building inspector, sent a chill of panic late Tuesday. Witry had been assigned to beef up a sandbag closure in the levee system on 9th Avenue. The news wasn't good. The sandbagging had not been done properly and the whole structure was weak. His assessment: the wall was going to fail. "I told him if this goes, the whole downtown goes," said Gaines. "I asked him what he needed. He told me the specifics and we had it there within minutes."

In addition to reinforcing the sandbag closure, Witry was also successful in building a secondary wall to contain the water if there was a breach. All through the night, Witry and a team of college students labored to reinforce the two walls. "It was an incredible effort," said Gaines. "They had such energy and it was such hard work. We are very grateful for their efforts. And that portion of the levee held."

More heavy rain was forecast for the night of June 10, and the predictions of two to four inches on the river basins did not disappoint. With

CEDAR RIVER: THE FIRST TO GO ✴ 97

Cedar Falls. COURTESY DENNY MILLS/GAYLON ISLEY

Tough work in Cedar Falls.
COURTESY SCOTT W. SMITH

Cedar Falls. COURTESY DENNY MILLS/GAYLON ISLEY

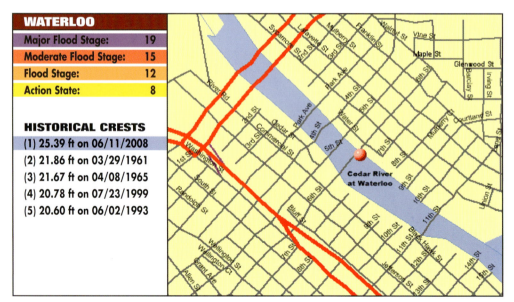

WATERLOO	
Major Flood Stage:	19
Moderate Flood Stage:	15
Flood Stage:	12
Action State:	8

HISTORICAL CRESTS
(1) 25.39 ft on 06/11/2008
(2) 21.86 ft on 03/29/1961
(3) 21.67 ft on 04/08/1965
(4) 20.78 ft on 07/23/1999
(5) 20.60 ft on 06/02/1993

COURTESY NATIONAL WEATHER SERVICE

no place to go, the floodwaters ballooned and, on June 11, the Cedar River crested in Cedar Falls at 102.13 feet, smashing the old record set in 1999 by six feet.

In Cedar Falls, the levee system held and the downtown was spared. However, north of the river in northern Cedar Falls, there was considerable flooding in what is considered part of the 100-year floodplain. Five hundred and eighty homes had water in them, and nearly half of those had water well into the first-floor living quarters.

THE POWER OF PREVENTION
Waterloo

The man responsible for keeping the city of Waterloo dry was Jamie Knutson. As the city's flood coordinator, his primary duties included maintaining the city's levee system. On June 1, the National Weather Service informed Knutson of the potential for

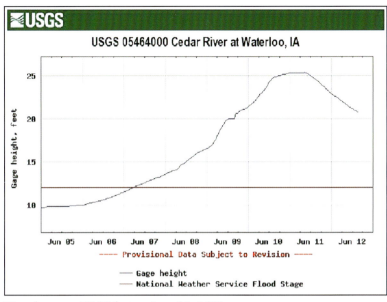
Waterloo crest 25.39 feet on June 11, 2008. COURTESY U.S. GEOLOGICAL SURVEY

Flood panels designed by the Army Corps of Engineers.
COURTESY WATERLOO CITY ENGINEERING DEPARTMENT

flooding on the Cedar River. The early projections predicted water levels to reach as high as they had been in 1993 and 1999. Knutson prepared accordingly, shutting off storm sewer outlets to the Cedar River to prevent water backing up into residences and businesses. Eight days later, every storm sewer outlet to the river was shut off, and floodgates on creeks that fed the Cedar River were closed. All flood prevention systems were in place and functioning as they should. It was the next day, on June 9, that things started to get dicey.

Water was coming up on the west side of town, with Dry Run Creek overflowing its banks. More water was rising on Blower Creek as it fed into the Cedar River near Evansdale. Even so, as of Tuesday, June 10, Knutson felt the town was holding its own against the rising tide, despite the fact that the 11th Street Bridge was closed.

On Wednesday "all hell broke loose," said Knutson. Rain the previous night was prolific, and with all the storm sewer outlets closed, the excess water had no place to go. "We were going into uncharted waters," said Knutson. Every employee of the engineering department was put to work erecting huge aluminum panels to hold back the impending flood. The panels were the brainchild of the Army Corps of Engineers. Designed in the 1970s, they were Waterloo's answer to the massive flooding it had experienced in the 1960s. The Army Corps of Engineers and the city of Waterloo anted up the funds, with the Corps putting up $44 million and the city $22 million. Each panel measured eight feet wide and three feet long. They were attached with clips to aluminum posts already planted in the ground. When installed, the panels run along the Cedar River for about eight blocks in the downtown area. For the past thirty years, they had been locked up in underground storage vaults. Until the floods of 2008, they had never seen the light of day.

"The panels were the key to saving the city," said Knutson. "We managed through the grace of God to get them in place in time. My gut tells me we were within three hours of having the whole downtown looking like Cedar Rapids," he added. In other words, much of Waterloo would have been submerged under several feet of water.

The panels were erected at the 11th Street Bridge, the Park Avenue Bridge, the 18th Street Bridge, and behind the art center, river plaza, and the Veteran's Memorial Hall.

The water coming up so quickly exhausted the manpower of city

Wet in Waterloo.
COURTESY WATERLOO CITY ENGINEERING DEPARTMENT

Downtown Waterloo.
COURTESY CANDY WELCH STREED

Union Pacific bridge—downtown Waterloo.
COURTESY KWWL SHOW AND TELL 7

employees. Knutson said whole neighborhoods stepped up, filling and stacking sandbags to protect their homes. "It was amazing to see everyone pull together and get everything done," said Knutson.

By Wednesday night, the rising water proved too much for the flood protection efforts on Blower Creek. "It overran everything that we had done," said Knutson. At that point, the west side of the downtown was under water. The only bridge open was Conger Street, and Knutson nearly lost that bridge when the road began flooding. For a while the Conger Street bridge was the only way to cross the downtown. Flooding continued from Evansdale to 18th Street and from the Cedar River to Franklin Street, encompassing the whole neighborhood in the southeast side of the downtown.

Amazingly, even with these two large sections flooded, only 5 percent of Waterloo's population was directly impacted by the flood. "It is a cool thing because of the levee system and what it was able to protect, and it was a spooky thing because residents don't have any idea of the impact of the flood," said Knutson. "We are one of two cities in the state that did not have levee failure," he added. "The bottom line is that the maintenance of the levee system was done, and done effectively, and it is why we survived. Water came up quickly but went down quickly. It was a combination of luck and maintenance. We didn't have breaches and we are very thankful for that. If a breach had occurred, our downtown would have looked like it did in the 1960s photos of flooding. It would have effectively wiped out hundreds of blocks of downtown Waterloo."

The day after the water crested in Waterloo, Knutson said a local engineering firm marked the high-water points and plugged the information into a current hydrological model. From an unofficial standpoint, the model determined that the city had experienced a 500-year flood.

When the crest was established in Waterloo on June 11, it was 3.5 feet higher than any previous flood. Hydrologist Jeff Zogg stated, "Once we saw what happened in Waterloo and Cedar Falls, the overall magnitude of the crest, we knew this was not just a record flood, this had the makings of a high-end historic event." He added, "What's more, we knew more rain was coming and it was only going to aggravate the situation."

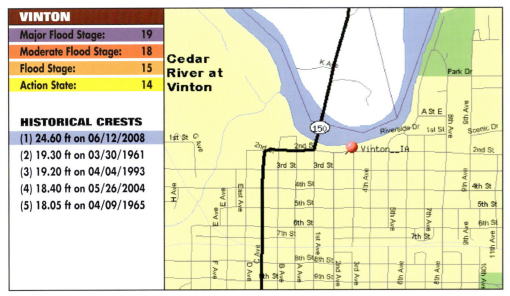

COURTESY NATIONAL WEATHER SERVICE

Benton County Law Enforcement Center. COURTESY LINDA MEEKER

ATTACK FROM BELOW

Vinton

Farther down on the Cedar, evacuations were ordered in parts of Vinton, a small farming community of about 5,000 that lies between Cedar Falls and Cedar Rapids. The Cedar River runs along the north side of town. The first crest projections seesawed back and forth. The first prediction set the crest at around twenty-one feet, which would top the previous high marks set in 1993 (19.2 feet) and 1961 (19.3 feet). On Monday, the projection was raised to between twenty-four and twenty-five feet. City leaders initiated sandbagging efforts, and hundreds of volunteers showed up. Sandbag levees went up around the electrical generation facility, fire station, and the Benton County Law Enforcement Center. On Tuesday afternoon, the crest projection was lowered back down to nineteen feet, but efforts continued. At four o'clock Wednesday morning, a frantic call came in to the sheriff department's communications center. "It was the National Weather Service trying to reach the emergency management manager. A mistake had been made. The crest projection was back up to twenty-four feet," said Benton County's Sheriff Randy Forsythe. The sandbagging notched upwards as volunteers launched a Herculean effort to save the power plant. One hundred Iowa National Guardsmen shoveled sand with teenagers, city rescue workers, and even inmates from the county jail. The sandbagging continued throughout the night, but the water came up too fast and too high. By Wednesday morning the floodwaters prevailed, surging underground and imploding beneath the floor of the power plant and flooding it. Power to the town was knocked out. "We lost four of the six generators," said Andy Lent, Vinton's city coordinator. Back-up generators were shipped in from as far away as Minnesota and South Dakota, and electricity was restored in just a day and a half. At the Benton County Courthouse, water backed up through a drain in the basement, where the Emergency Management office is located. In all, fifteen city blocks were covered by floodwaters, and 80 to 100 homes were affected.

The other big story was the Benton County Law Enforcement Center.

Slow going. COURTESY LINDA MEEKER   **Yield to the flood.** COURTESY LINDA MEEKER   **Bagging with the Guard.** COURTESY LINDA MEEKER

Sheriff Forsythe, a thirty-two-year veteran with the Benton County Sheriff's Department, made the decision early to evacuate the administrative side of the building. While operations continued at the communications center and jail, the sheriff ordered a semitruck and loaded it with everything that wasn't bolted down. When the river crest was projected upward for the second time, the entire building was sandbagged. By late Wednesday, Sheriff Forsythe told staffers to prepare the inmates for evacuation. They were fed dinner and taken by boat to a waiting bus, to be transported to neighboring jails. "I don't think they understood the magnitude of the flooding until they stepped outside," said Forsythe. "We actually had a couple of inmates who didn't want to leave but asked to stay to continue sandbagging. Of course we couldn't do that."

That night, water crept up to the top of the levee. Pumps were installed to keep up with the seepage. Three volunteers took the night watch. At midnight, heavy rains appeared and quickly overwhelmed the pumps. At three in the morning, a large ten-inch pump was brought in by end loader, and quickly began to pump the water back down. For about an hour, the pump kept pace with the floodwater until exactly one minute before four, when the dike failed. "It broke in the northeast corner of the building. There was a lot of water moving down the alley and it got to swirling and hitting the sandbag wall. The particular area where it went was the lowest part of the dike," said Forsythe.

The inside of the building had thirty-two to thirty-four inches of water. The emergency radio equipment was damaged, along with the infrastructure of the building. Damage was estimated to be around $350,000.

"The flooding basically ruins everything because of the humidity it creates in the rooms and the chemicals in the water," said Forsythe. "Computers that weren't even close to the water were ruined because of the humidity or the power surge when the water hit. I have been here for thirty-five years and have never seen anything like this. The flooding surpassed every record by at least four and a half feet."

On June 12, the river reached its high point in Vinton, an all-time record crest of 24.60 feet.

THE LOST CITY
Palo

Downstream from Vinton, the 950 residents of Palo awaited their fate. In a morning briefing with city and county officials, Governor Chet Culver said, "It is a very, very serious situation," adding, "We haven't seen anything like this in a long, long time."

Palo is a community where families tend to stay and grow. "We have whole families who live here," said Paula Gunter, mayor pro tem. "Grandparents, parents, and their children growing up and raising families."

The Cedar River flows on the east side of town. Two creeks also wind through, eventually emptying into the Cedar. Despite playing host to a large river and two creeks, only about 15 percent of the town flooded when the Cedar crested at a near record of 19.2 feet in 1993.

When city leaders began to prepare for high water on June 8, they assumed the water would again reach about nineteen feet. "We brought in thirty loads of sand and set up a Linn County emergency management post in town," said Gunter. A voluntary evacuation was issued for the southeast section of town.

Early Tuesday morning, officials scheduled emergency meetings for every four hours, and a mandatory evacuation order was issued for half the town. Late Tuesday, it became clear the expected nineteen-foot crest on the Cedar was going to be higher, much higher. "Every time we had our emergency meetings, the levels kept rising, first by six to eight inches at a time, and then it became feet at a time. Then they told us to plan for twenty-four to twenty-six feet," said Gunter.

Tuesday afternoon, the river started creeping toward town, first overtaking the strawberry fields and then inching to the town's main thoroughfare, Blairs Ferry Road. "In the beginning we were in denial, but as we sat in those emergency meetings we realized we really needed to save the infrastructure of the town and save people's homes," said Gunter. "The biggest responsibility that we had was to keep people safe," she added.

PHOTOS THIS PAGE: Palo succumbs to the Cedar River. COURTESY PALO FIRE DEPARTMENT

Palo. COURTESY HANDS ON DISASTER RESPONSE

Palo. COURTESY PALO FIRE DEPARTMENT

Palo. COURTESY HANDS ON DISASTER RESPONSE

Town residents were also not taking any chances, and began preparing for the worst. Trailers and trucks of every type poured into town as people packed up the contents of their homes. "I have never seen so many snowmobile, livestock, and equipment trailers in my life," said Gunter. "Several people even ripped up the carpets in their basements, knowing it would be easier to clean if the carpets weren't there."

Meanwhile, volunteers pushed themselves hard to save the town. City officials estimated approximately 1,200 volunteers filled 80,000 sandbags.

On Tuesday night and into Wednesday morning, the Cedar River rose so high the two creeks through town began to back up and overflow their banks. Once the water from the Cedar overtopped Blairs Ferry Road, a huge wave gushed into town, flooding businesses and homes with eight inches of water in just the first hour. Late Thursday night and into the morning of Friday the 13th brought more heartache. More than six inches of rain fell, creating flash flooding and adding to the nightmare. By the time the Cedar River crested at 31.12 feet, the entire town was covered in water, ranging from two to eleven feet deep. "That day the fire department and the National Guard were doing water rescues . . . bringing people out by boats and by two-and-a-half-ton Army trucks. About eighteen families had to be rescued," said Gunter.

In a town as tightly knit as Palo, the devastation was felt throughout families. "When something like this happens, you go to your best resource, your family. But we didn't have that because whole families were affected here," said Gunter. All but 10 of Palo's 423 homes and businesses were flooded, and all 950 residents were evacuated for four days. "But the really important thing here is that we didn't lose a single life. And with that much water it could have very easily happened."

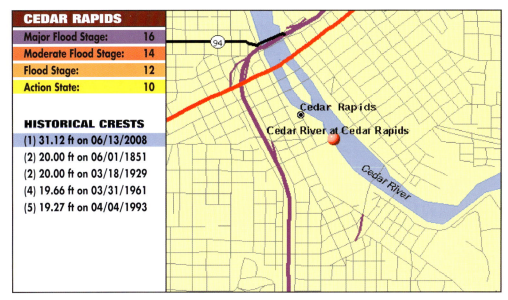

COURTESY NATIONAL WEATHER SERVICE

SPECIAL CREDIT

The Flood of 2008 has triggered an outpouring of support from the community. One of those giving back is Andrea Lynn Photography. Andrea Lynn Photography has donated the usage rights to the following photographs to raise money for charity and is donating 100 percent of the profits received from flood photograph sales in 2008 to flood recovery in addition to working with and donating her photographs to multiple Iowa-based charities.

THE UNTHINKABLE
Cedar Rapids

As early as Sunday, June 8, Cedar Rapids was aware that a record flood was on its way. However, expectations were that the crest would be similar to the 1929 stage of twenty feet. By 11:45 A.M. Tuesday, the number was adjusted upward to twenty-two feet—two feet higher than the river had ever gone in Cedar Rapids history. It was the level of a 100-year flood.

To hold off the water, city officials built two temporary dikes to provide protection to the Time Check neighborhood in northwest Cedar Rapids and homes near the Czech Village, and another to protect Osborn Park and the Sinclair packinghouse. The dirt for the dikes was transported from the Eastern Iowa Airport. Additional dirt was used to increase the height of the permanent Time Check neighborhood levee. Evacuations were recommended for residents of about a dozen homes on Ellis Road NW.

Concerns were raised that the CRANDIC railroad bridge downstream of 8th Avenue would be swept away. Governor Culver was informed by engineers that there was a chance the bridge would float off its piers if the water got any higher than expected. City officials estimated over 100 employees, some working sixteen-hour shifts, were battling the rising water with 30,000 sandbags in place and 100,000 empty bags ready to be filled. As concerns mounted, the city of Cedar Rapids set up a twenty-four-hour nonemergency help line to assist citizens with flood-related questions.

On Wednesday, June 11, following another night of rain on the headwaters of the Cedar and Iowa rivers, crest predictions were increased again for places such as Cedar Rapids and Iowa City. Rumblings of a flood seen just once every 500 years were now being heard around the state.

In Cedar Rapids, a revised crest of 24.5 feet was expected in less than two days. The benchmark level of a 500-year flood was 26.5 feet. When news of the 24.5 height reached city engineers, they knew it was the beginning of the end. There would not be sufficient time to increase the height of levees and dikes. Sandbags could never be filled and placed. Cedar Rapids was facing a crippling blow and was at the mercy of the Cedar River. The city was in full crisis.

Water would soon flow onto the 1st Avenue Bridge deck in downtown Cedar Rapids. That structure and the city's three other main

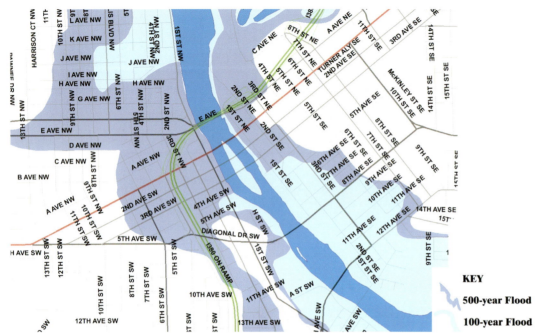

100-year/500-year floodplain for Cedar Rapids.
COURTESY LINN COUNTY EMERGENCY MANAGEMENT

KEY
500-year Flood
100-year Flood

FLOOD STATEMENT
NATIONAL WEATHER SERVICE QUAD CITIES IA IL
1046 AM CDT WED JUN 11 2008

...A HISTORIC HYDROLOGIC EVENT IS EVOLVING...
UNPRECEDENTED RIVER CRESTS ARE EXPECTED...

...RECORD FLOODING IS EXPECTED ON THE CEDAR AND IOWA RIVERS IN EASTERN IOWA. THIS IS A SERIOUS SITUATION.

COURTESY NWS STORM PREDICTION CENTER

downtown bridge spans were already crowded with sightseers taking pictures of the boiling water inches below. The bridges were ordered closed and the crowds sent packing. The 8th Avenue railroad bridge was also shut down, and hopper cars weighted down with rock spanned the structure in hopes of keeping the bridge from being washed off its piers.

In flood-prone areas, Alliant Energy cut gas service to several thousand customers. Electrical power to the heart of the city was also likely to be severed when the river reached its projected levels. City crews put sewer pipes and sandbags around sewer openings along the river, to contain water backing up from sewer outlets. At the Public Works facility, volunteers filled sandbags around the clock as the American Red Cross staffed a shelter on the city's northwest side.

June 12 was the most insane day of Jeff Zogg's career as a hydrologist. While the crest was bearing down on Cedar Rapids, rain was falling and more was expected into the night. Crest levels were going up again. With every river in the eastern half of Iowa in flood, he and his office were in constant contact with the U.S. Geological Survey, getting updated flow measurements. Conference calls were ongoing with county, state, and federal agencies. Requests from the media were being addressed. On top of that, severe afternoon and evening thunderstorms required numerous warnings and attention. Zogg's office was abuzz while his personal focus turned to the catastrophic situation developing in Cedar Rapids.

With up to five inches of new rain falling on top of the approaching crest in Cedar Rapids, flash flooding was occurring on top of the existing flooding. It was absolutely the worst-case scenario imaginable. It was also making crest predictions difficult, as the freshwater overwhelmed previous model estimates. Zogg said, "It was increasingly difficult to grasp the magnitude of the flood. It just blew my mind away."

With the rivers now in uncharted territory, no one knew how high they would go. Topography dictates how the water spreads, and

Cedar Rapids. COURTESY ANDREA LYNN PHOTOGRAPHY Cedar Rapids. COURTESY ANDREA LYNN PHOTOGRAPHY Cedar Rapids. COURTESY ANDREA LYNN PHOTOGRAPHY

experts can make approximations but, until it happens, nobody knows for sure. A team of experts was assembled to get a handle on the crest estimates, including members of the USGS and the hydrology lab in Washington, D.C. It was such a challenge and so essential that Zogg said, "We were talking to anyone with experience and knowledge of this sort of situation."

Adding to the difficulties was the fact that the river was now going to top the bridges in downtown Cedar Rapids. The bridges would act like dams and cause the water level to jump upstream of the obstruction. Zogg likened the effect to driving a car at seventy miles per hour and then hitting the brakes. "The back end of the car goes up—and water does a similar thing when it hits a bridge." He added, "It could raise river stages upstream by three to five feet." Following the event, Zogg indicated the bridges in Cedar Rapids helped raise the crest, but by how much he didn't know, saying, "It was a factor, but the flash flood and the sheer volume of the water had a bigger effect."

Another significant issue hydrologists and emergency officials faced was a key river gauge that was lost in Cedar Rapids due to a power outage Wednesday night. The backup batteries failed and the river's stage was unknown as the crest approached. Through most of Thursday, June 12, the river was projected to crest at 24.7 feet. A U.S. Geological Survey team eventually restored the gauge, and when the new data was analyzed by computer models, it was astonishing. At 3:45 P.M., the river had already soared to 28.5 feet; it was not expected to crest until Friday morning, June 13, at thirty-two feet.

For the city of Cedar Rapids, the news was like hearing a death sentence. Not only would the river be twenty feet above flood stage, it would be twelve feet higher than any other flood in 157 years. The crest would be more than five feet above the projected 500-year flood height. The goals of emergency management shifted from protection to survival.

By late Thursday evening, Alliant Energy reported 15,000 customers were without power. The utility's namesake tower was evacuated earlier in the day, and the computers used to monitor generating and distribution systems were no longer available. Residents were told that outages could last more than a week.

City officials estimated that at least 3,200 homes had gone under

THE FLOOD WARNING CONTINUES FOR THE CEDAR RIVER AT CEDAR RAPIDS.

* UNTIL FURTHER NOTICE.
* AT 3:45 PM THURSDAY THE STAGE WAS 28.5 FEET.
* FLOODING IS OCCURRING AND RECORD FLOODING IS FORECAST.
* FLOOD STAGE IS 12 FEET.
* FORECAST...RISE ABOVE FLOOD STAGE TODAY...AND CONTINUE RISING TO 32.0 FEET FRIDAY MORNING. THE CREST FORECAST HAS BEEN ADJUSTED UPWARD DUE TO THE COMBINATION OF LOCAL RUNOFF FROM TODAY'S HEAVY RAINFALL...ALONG WITH THE COLLAPSE OF THE RAILROAD BRIDGE ON THE SOUTHEAST SIDE OF TOWN. THE COLLAPSE OF THE RAILROAD BRIDGE WILL LEAD TO HIGHER STAGES THAN WOULD OTHERWISE BE EXPECTED.

COURTESY NWS STORM PREDICTION CENTER

water, along with several hundred businesses. Many more had flooded basements. At least 8,000 people had been evacuated from their homes.

On May's Island, the Linn County Courthouse and Linn County jail were submerged. The inmates were evacuated by bus Wednesday morning. City Hall was also taking on water, as well as other historical landmarks throughout the city. One of the most notable was the refurbished Paramount Theatre and its rare Wurlitzer organ.

With water seemingly everywhere, one of the great ironies of the flood was its ability to jeopardize the freshwater supply of Cedar Rapids. As the situation grew worse, the city suddenly found itself running out of freshwater. According to Craig Hanson, Cedar Rapids public works manager, three of the four lateral water wells that produce eight to twelve million gallons of water per day had stopped functioning. Additionally, forty-six single-shaft wells that produce one million gallons per day were also flooded or out of power, reflecting the urgency of protecting the last well.

Despite its best efforts, the city was losing the battle. In one last heroic effort, Hanson ordered fifty tons of sand, trucks full of empty sandbags, ties, and shovels moved to the site. He then called a local TV station and

Cedar Rapids. COURTESY ANDREA LYNN PHOTOGRAPHY   Cedar Rapids. COURTESY ANDREA LYNN PHOTOGRAPHY   Cedar Rapids. COURTESY ANDREA LYNN PHOTOGRAPHY

Cedar Rapids. COURTESY ANDREA LYNN PHOTOGRAPHY Cedar Rapids. COURTESY ANDREA LYNN PHOTOGRAPHY Cedar Rapids. COURTESY ANDREA LYNN PHOTOGRAPHY

put out the word that the city needed help. In no time, 1,000 volunteers showed up to fill and place 8,000 sandbags around the well.

With sandbags and generators in place, the well continued to function, although the remaining water supply was pumped at just 25 percent capacity. To ensure enough water for firefighters and the basic needs of residents, most of Cedar Rapids' major businesses voluntarily shut down, and residents were begged to conserve.

In addition, all of the bridges in the downtown area were closed, along with block after block of streets. Two lanes of a jammed Interstate 380 were the only way to get from one side of the city to the other. Downstream, the Cedar had also closed Iowa's main east-west thoroughfare, Interstate 80 south of Tipton. Many smaller bridges and roads were also shut down, keeping the Iowa Department of Transportation busy posting detour signs.

When the clock struck midnight on June 13, the Cedar River hovered around thirty-one feet. Cedar Rapids resembled a vacuous black hole, with power and lights cut throughout the heart of the city. At Mercy Hospital, on the southeast side, a dramatic scene unfolded when water invaded the hospital. Patients and staff were forced to evacuate in the middle of the night for higher ground. The hospital was so far out of the 500-year floodplain that officials had not planned on making the move, especially since the water was not expected to go over twenty-five feet for much of the day. The evacuations were orderly and went off without a hitch.

Finally, at 10:15 A.M. June 13, the Cedar stopped climbing in Cedar Rapids. The gauge had leveled off at an all-time record of 31.1 feet. The epic crest had laid waste to much of the city and its infrastructure. Throughout the day, television sets and newspapers all over the world showed the plight of Cedar Rapids. With the water going down, the national media would soon leave town, but for the citizens of Cedar Rapids the legacy of the Great Flood of 2008 would only grow bigger with time. ✸

3:28 PM CDT FRI JUN 13 2008
...FLOOD WARNING REMAINS IN EFFECT UNTIL FURTHER NOTICE...

THE FLOOD WARNING CONTINUES FOR
THE CEDAR RIVER AT CEDAR RAPIDS.
✸ UNTIL FURTHER NOTICE.
✸ AT 3:00 PM FRIDAY THE STAGE WAS 31.0 FEET... AND FALLING.
✸ RECORD FLOODING IS OCCURRING.
✸ FLOOD STAGE IS 12 FEET.
✸ FORECAST...FALL TO 28.5 FEET SATURDAY MORNING

COURTESY NWS STORM PREDICTION CENTER

CREDITS LEFT TO RIGHT: UNIVERSITY OF IOWA, CITY OF CORALVILLE, UNIVERSITY OF IOWA, UNIVERSITY OF IOWA, UNIVERSITY OF IOWA

IOWA RIVER: WORSE THAN '93

Iowa City

As the crow flies, Cedar Rapids and Iowa City are only about twenty-five miles apart. Both suffered hard knocks from the floods of 2008. Despite similar scenes of devastation and heartbreak, the Iowa River took its toll in a managed way when compared to the raging torrents of the Cedar River. This was due to the fact that upstream from Iowa City is a dam so large that its forty-one-mile-long reservoir maintains the runoff of 3,000 square miles of land above it. Designed by the Army Corps of Engineers in 1938, the primary purpose of the Coralville Dam is to manage the Iowa River and provide flood protection to 1,700 square miles of river valley below.

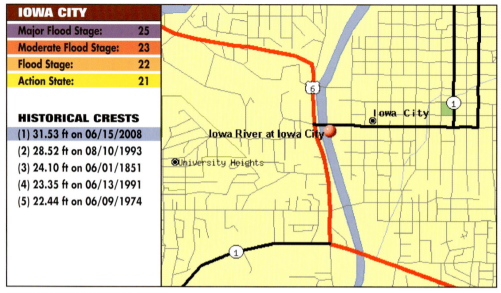

COURTESY NATIONAL WEATHER SERVICE

The "Coralville Strip"—closed eastbound to Iowa City. COURTESY CITY OF CORALVILLE

For the most part, the dam has been effective in achieving its goals. However, when the forces of nature act as they did in 1993 and 2008, even dams have limitations. During times of high water, the 350-foot-long, 23-foot-wide conduit outlet can release only so much water without flooding the valley below. In both 1993 and 2008, the intake of the lake was greater than the amount discharged. Eventually, the water level rose to the top of the 712-foot spillway and cascaded into the river valley below.

When flooding is anticipated, the Corps' goal is to manage the discharge so that an acceptable balance is maintained on both sides of the dam for as long as possible. Discharge levels are altered as events unfold. For each increase in discharge, there is a corresponding increase in river levels in places such as Coralville and Iowa City. In this way, a flood can be managed—until the lake exceeds the spillway. At that point, alterations can still be made in the amount of water discharged, but control is more in the hands of nature than those of humans.

Due to the same weather parameters that produced the historic flooding on the Cedar, the Army Corps of Engineers had been steadily increasing the discharge of water below the spillway since June 5. On June 9, at 10:47 A.M., Army Corps of Engineers staff informed Iowa City that due to heavy rainfall locally as well as in the northern watershed on Sunday night, the outflows from the reservoir would need to be increased to 21,000 cubic feet per second by Wednesday, June 11. Even with the increased discharge, floodwaters were expected to flow over the Coralville Dam spillway around June 11, a feat that had occurred only once before during its fifty-year history. The Iowa River was headed toward historic levels, and Iowa City residents were told to plan for the worst.

An unknown variable impacting Iowa City's crest was a cofferdam located near the Iowa Memorial Union. The barrier was built by the University of Iowa so crews could install two thirty-six-inch underwater pipes connecting the university's east and west chilled water plants. The cofferdam was such an imposing obstacle it was elevating water levels upstream as far away as Coralville. With the

100-year/500-year floodplain for Iowa City.
COURTESY IOWA CITY PUBLIC WORKS DEPARTMENT

KEY
500-year Flood
100-year Flood

cofferdam in place and water projected to go over the spillway, officials warned the flood would be worse than 1993 and could actually approach 500-year levels.

In response to the dire warnings, the Johnson County Sheriff's Department banned all watercraft on the Iowa River. Free sand and sandbags were offered at three distribution points. Several roads in Iowa City and Coralville were closed, as was the University of Iowa's Art Campus.

Iowa governor Chet Culver indicated that the local governments of eastern Iowa were as prepared as could be expected for flooding unlike any in living memory, and the state was ready to support them. Culver added, "It's going to be in all likelihood a very, very difficult few days and perhaps weeks ahead."

By 5:00 P.M. June 10, water trickled over the spillway of the Coralville Dam; by 9:00 P.M. it had built to a steady flow. Above

COURTESY DARYL SCHEPANSKI   Two views of water flowing over the spillway of the Coralville reservoir.   COURTESY CITY OF CORALVILLE

1046 AM CDT WED JUN 11 2008
...FLOOD WARNING REMAINS IN EFFECT UNTIL FURTHER NOTICE...

THE FLOOD WARNING CONTINUES FOR
THE IOWA RIVER AT IOWA CITY.
* UNTIL FURTHER NOTICE.
* AT 10:00 AM WEDNESDAY THE STAGE WAS 25.9 FEET...
  AND RISING.
* MAJOR FLOODING IS OCCURRING.
* FLOOD STAGE IS 22 FEET.
* FORECAST...CREST AROUND 30.9 FEET NEXT THURSDAY.
* IMPACT...AT 29 FEET...SERIOUS FLOOD DAMAGE OCCURS
  AT THE UNIVERSITY OF IOWA CAMPUS.

COURTESY NWS STORM PREDICTION CENTER

Kinnick Stadium—home of the Iowa Hawkeyes. COURTESY FARSHID ASSASSI, ASSASSI PRODUCTIONS AND THE ARCHITECTS: NEUMANN MONSON/HNTB

the dam, water entered the lake at 21,000 cubic feet per second while, below it, with the discharge gates wide open, water exited at 20,000 cubic feet per second. It was estimated by architects from Neumann Monson that water surging at 20,000 cubic feet per second would fill Kinnick Stadium (home of the Iowa Hawkeye football team) in less than seventeen minutes if the stadium's corners were enclosed. (On June 15, peak outflow reached 39,500 cfs, and Neumann Monson estimated that at that rate the stadium would have been filled in 8.4 minutes.)

Meanwhile, up to one million sandbags had been put in place in Johnson County, home to Coralville and Iowa City. Evaluations were also underway regarding the need for evacuations in certain neighborhoods such as those near Normandy Drive west of City Park. The dilemma was that current city guidelines did not give city officials the authority to order such evacuations.

Following another stinging night of heavy upstream rains, on the morning of June 12, Iowa City took the news of a crest increase as just another challenge. With the river now expected to rise to thirty-one feet, the University of Iowa assessed the vulnerability of campus buildings. Previously, buildings had been protected based on the measurement of the 100-year flood level plus one foot. Officials now decided to add an additional eighteen inches to that equation. Faculty members were informed that they should pack to be out of their offices until August.

With the likelihood of residential evacuations in coming days, the city council of Iowa City voted 6-0 to give the mayor the power to order evacuations and take "any other extraordinary measures" to preserve life or respond to a disaster. Violating an order would be illegal, punishable as a simple misdemeanor. Mandatory evacuations were already imposed on Edgewood Drive and Shadow Lane in Coralville.

Meanwhile, 640 Iowa National Guard members were activated for statewide flood duty, and another 1,700 placed on alert.

Often, the difference between staying high and dry or being submerged in several feet of floodwater is the strength of a sandbag wall or a levee made of dirt. On Thursday, June 12, both Coralville and Iowa City saw heartbreaking examples of the latter.

In the Normandy Drive neighborhood, Iowa City police evacuated residents at 2:00 A.M. due to concerns over the rapidly rising waters and the strength of a four- to five-foot sandbag levee protecting dozens of homes. Volunteers had worked around the clock for over a week to protect the neighborhood, but the water proved too much. Sandbag protection around a sewer intake ultimately failed around 10:00 A.M., and the Normandy neighborhood was quickly submerged.

Back in Coralville, the combined water pressure of Clear Creek and the Iowa River forced gaping holes in the embankment of the CRANDIC rail-

Determination. COURTESY CITY OF CORALVILLE

Living on the Iowa River—literally. COURTESY UNIVERSITY OF IOWA

The Cofferdam—a flooding enhancer?
COURTESY UNIVERSITY OF IOWA

road line near Old Chicago. Before the water completely broke through, the city was able to provide about three to four hours notice that the Coralville strip would flood. Power to the area was cut by Mid American Energy, and by evening the only way to navigate the strip was by boat.

At the University of Iowa, seventeen buildings on the campus were closed and two more were planning for evacuations. The nineteen buildings represented about 12 percent of the campus's square footage. Massive sandbag walls lined both sides of the river in hopes of protecting the endangered buildings.

Outflow from the Coralville Dam was now at 26,000 cubic feet per second, and the river was five feet above flood stage in Iowa City, closing in on the record crest of 28.5 feet set in 1993. Residential evacuations were ordered on all or parts of three dozen city streets. At 8:00 P.M., Interstate 80 was closed east of Iowa City. The ensuing detours added 110 miles to a trip from Des Moines to the Quad Cities, and an additional 2.5 hours.

As luck would have it, Friday the 13th dawned bright and sunny in Iowa City. Unfortunately, the Iowa River was pushing thirty feet and was at the highest level in city history. With the crest several days away, about 1,500 residents had been displaced from at least 500 homes. Another 500 citizens of Coralville were homeless, according to emergency shelters there.

With the water climbing higher, more and more roads were going under, leading to major traffic woes. Officials advised patience as police tried to fit the same number of cars onto one road that were usually spread out among five. The problems were expected to only grow worse as the Interstate 380 corridor closed between Iowa City and Cedar Rapids at 6:00 P.M. Iowa City's downtown bridges were also in danger of closing, which would section the city into two distinct halves, west and east. The police and fire departments were making provisions to have proper staffing and equipment on both sides of the river in case the bridges needed to be closed.

University of Iowa officials continued their scramble to evacuate buildings and adjust flood prevention efforts. With a thirty-three-foot crest anticipated, severe flood damage was a certainty—and in some cases a reality—as several campus buildings were already taking on water.

The "Coralville strip" was now beneath five feet of mud-laden water. The tree-lined thoroughfare so familiar to Iowa Hawkeye fans was dotted with boats instead of its usual array of cars.

The morning of June 14, the Army Corps of Engineers informed Iowa City officials that projections made on June 13 remained consistent, with no significant change. The reservoir was expected to crest at 717.75 feet on Monday, June 16, with an outflow of 42,500 cubic feet per second.

Cleaning the campus.
COURTESY UNIVERSITY OF IOWA

Walls of sand protect the Iowa Memorial Union.
COURTESY UNIVERSITY OF IOWA

June 14—volunteers converge.
COURTESY UNIVERSITY OF IOWA

Water was being discharged out of the reservoir at 39,000 cubic feet per second, smashing the old mark of 28,000 cubic feet per second set in 1993. The river would likely climb another 2.5 feet; the crest in Iowa City was expected to reach thirty-three feet the afternoon of June 16.

Back on the campus of the state's oldest university, up to 2,000 volunteers piled sandbags in the fight to save twenty threatened buildings. For the most part, the river was winning the battle. Sandbagging efforts were suspended by late afternoon because of severe thunderstorms. By then, water was pouring into twelve campus buildings. The university's power plant was also shut down due to water infiltration.

As evening arrived, water lapped at the pavement of the Burlington Street Bridge, but both it and the Benton Street Bridge remained open. Had the eventual crest been five inches higher, both bridges would have been closed. As it was, traffic was reduced to one lane in each direction, and at times motorists were forced to drive through several inches of water.

Unexpectedly, June 15, 2008, turned out to be a day of mild celebration. To everyone's surprise, the Iowa River was on its way down. The good news was conveyed by the Army Corps of Engineers in a morning briefing that proclaimed Coralville Lake had crested the previous evening (Saturday, June 14) at 717.0 feet, five feet over the spillway. Downstream, the ensuing crest of 31.53 feet occurred in Iowa City at 6:30 A.M., lower than the anticipated crest of thirty-three feet.

Officials were quick to point out that the flood would still be a prolonged event. Flooding from rivers, such as the Cedar River in Cedar Rapids, subsides much quicker than flooding associated with the draining of a reservoir. Army Corps officials stated that if conditions remained dry, outflows would gradually fall over the next week, and that it was likely water would be below the spillway in seven to nine days.

For residents of Iowa City, Coralville, and the University of Iowa, it was time to take a deep breath and switch gears. The Flood of 2008 had moved from a level of protection to one of recovery. Hancher Auditorium and the University of Iowa Arts Campus were still under several feet of water, as were several residential neighborhoods. Sandbags—1.4 million of them—stood stoically around hundreds of homes, businesses, and roads. Inch by inch, as the water began to recede, the scope and scars of the 500-year flood were revealed. With exposure came the anguish and despair that disasters bring. On the other hand, the determination and spirit that was evident during the flood continued to shine during the recovery. It would take time and effort, but brighter days were on the way.

COURTESY BOB PATTON, *IOWA CITY PRESS-CITIZEN*

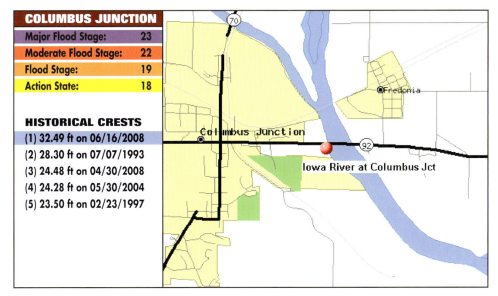

COURTESY NATIONAL WEATHER SERVICE

COME HELL OR HIGH WATER
Columbus Junction

What goes up must eventually come down, and downstream of Iowa City in the town of Columbus Junction, longtime city council member and mayor pro tem Mark Huston was the first to see how high the crest was expected to rise on the Iowa River. It was Sunday, June 8, and the National Weather Service was predicting the Iowa River to crest at 29.7 feet, which was nearly a foot and a half higher than the record crest set in 1993.

Galvanized into early action, city leaders called an emergency meeting for the following morning. The first order of business was to decide what could and could not feasibly be saved. The decision was made to save the downtown businesses at all costs and to allow the floodwaters to encompass the county fairgrounds.

Columbus Junction. COURTESY DON BEAN

Columbus Junction. COURTESY JOSEPH L. MURPHY, IOWA FARM BUREAU

A call went out for help, and by Tuesday morning, June 10, forty to fifty trucks, 200 to 300 volunteers, and various pieces of heavy equipment were all ready to begin work. "We put the first load of sand down at 10:00 that morning," said Hal Prior, a Columbus Junction city council member. The work grew more intense when news came that the crest had been raised to thirty-two feet. By the end of the day, 15,000 sandbags had been filled and half of the temporary levee was built along Highway 92.

By midafternoon the next day, work was proceeding well ahead of schedule. It wasn't until 6:30 that evening that the first alarm was sounded. The water was rising at a faster rate than expected, and possibly could start pouring over the old levee. Undaunted, city leaders shut down Highway 92, brought in portable lights, snacks, and bottled water, and worked until 10:00 that night securing the levee.

On Thursday, travel was a nightmare. With traffic stopped on the highway, it took nearly four hours to find a passable detour that crossed the river. Some volunteers, trapped on the wrong side of the river, resigned themselves to spending a few nights away from home. Late Thursday, leaks appeared in the levee and work started to reinforce weak areas. Meanwhile, a second levee wall was being built around the town's water treatment plant. To add to the tension, a severe storm broke loose that evening, producing a tornado that touched down just a few miles south of town. The storm also dumped another four inches of unwelcome rain, and the flooding grew worse. National guardsmen came in to help. They plowed through floodwaters reaching halfway up the doors of their trucks as they transported people back and forth across the levee. That night, the water approached twenty-eight feet.

Friday the 13th marked the breaking point. Predictions for the crest were now running as high as thirty-six feet. The command post was moved to higher ground, and all people and businesses were evacuated out of the low-lying areas. Only essential people were allowed in the areas where the water was expected.

Hal Prior was telephoned at 3:30 on Saturday morning. An emergency meeting had been called, could he come right down to the command post? When Prior arrived, the mayor and vice mayor were both on the phone to state officials. The crest was now predicted for

Columbus Junction. COURTESY TAMMY K. VIRZI

Columbus Junction. COURTESY TAMMY K. VIRZI

thirty-eight feet, and officials in Des Moines promised more help. At 4:30 in the morning, a code red was issued for 150,000 more sandbags, and an hour and a half later help arrived. "It was an awesome sight. More than 200 volunteers showed up in the pre-dawn hours to help sandbag," said Prior. About midmorning, the water started coming through the railroad tracks, and city leaders realized they couldn't stop it. Efforts turned to saving the water treatment plant. Volunteers with ropes and lifelines worked to plug holes in the sandbag walls, but the job grew too dangerous and the decision was made to shut down the plant and seal it off.

At 6:30 A.M., the river reached thirty-two feet and the south end of the levee started to give, and then failed. Water began to flow like an incoming tide and in just minutes started to fill up the downtown. "People were crying. They didn't want to stop sandbagging. They had worked so hard," said Prior.

In the end, up to twelve feet of water inundated the lower reaches of the downtown. But because of the work and sacrifice of so many, there were no injuries, and all businesses and homes affected were able to evacuate everything of value. Only the structures themselves sustained any damage.

That Sunday a free community dinner was held. It was an opportunity for Mayor Dan Wilson to thank the many who had given so much when their town needed them. City officials also had good news. A 3,000-foot fire hose had been tapped into the Tyson Foods water supply and would provide water for townspeople until their own plant was repaired.

While plans for recovery began, plans of another kind were in the works. Danielle Ritter called city hall and asked how she could marry her fiancé, who was a national guardsman sent to Columbus Junction. She told city leaders she was going to get married come hell or high water. With plenty of high water, Columbus Junction embraced her wish and threw themselves into planning a wedding. "Their wedding had been postponed several times and we wanted to do something positive," said Prior. "So we arranged for the couple to get married on Highway 92, which was still closed because of the floodwaters." So with a backdrop of murky brown water, Iowa National Guard Spec. Curtis White and Danielle Ritter, both of Macksburg, Iowa, were joined as husband and wife. ✸

CREDITS LEFT TO RIGHT: LUTHER COLLEGE, SEELEY PHOTOGRAPHY, SEELEY PHOTOGRAPHY, LUTHER COLLEGE, LUTHER COLLEGE

NOTEWORTHY MENTIONS

ADDING INSULT TO INJURY: THE ONE-TWO PUNCH
New Hartford

While New Hartford, Iowa, isn't exactly on the beaten path, its 659 residents have always enjoyed a peaceful and productive existence. That is, until the late spring of 2008, when the forces of nature unleashed their fury on this unassuming town. Just two weeks after a violent EF5 tornado had taken the lives of two citizens, a roaring flash flood claimed the rest of town.

New Hartford. COURTESY DENNY MILLS AND GAYLON ISLEY

Elkader. COURTESY KWWL SHOW AND TELL 7

According to city officials, everyone who was spared by the tornado was hit hard by the rain-swollen Beaver Creek. The floodwaters swirled into New Hartford after topping a dike west of town on the afternoon of Sunday, June 8. While many residents had already been evacuated, about 150 had to be rescued by firefighters in boats. Some people trapped in cars were rescued by human chains up to fifteen people long.

Fire Chief Brad Schipper, who ended up spending the night in the swamped fire station, indicated that only a handful of the city's homes were not flooded. Coming on the heels of the recent tornado, the untimely flooding added another major challenge to the shell-shocked town of New Hartford.

GOBBLED UP BY THE TURKEY

Elkader

The scenic Turkey River, with its centerpiece Keystone Arch Bridge, is one of Elkader's main attractions. This river in northeastern Iowa can also be a source of worry and concern. In the days leading up to the river's crest on June 9, residents braced themselves for the unknown as the water rose to new record heights.

On the south side of town, residents prayed that their levee, built in 1997, would protect them. But those prayers went unanswered. When the Turkey crested at 30.9 feet (smashing the 2003 record of 27.3 feet), the water simply flowed over the top. "There was a lot of disbelief," said city administrator Jennifer Cowsert. Four city blocks in the downtown were also flooded. In a town of not quite 1,500 residents, fifty-three homes and twenty-eight businesses were impacted.

Also destroyed in the flood were two pipes that ran along the river bottom. They were used for the town's water supply and its wastewater treatment facility. "The force of the flooding broke the pipes," said Cowsert. "So now we are going to have to dig deeper and bore under the river and through bedrock to replace them."

People from the community, nearby towns, and neighboring states came to assist Elkader. It was all part of the spirit that founded this small Iowa town in 1846 when city leaders named it after an Algerian hero named Abd el-Kader.

Anamosa. COURTESY SEELEY PHOTOGRAPHY

Anamosa. COURTESY SEELEY PHOTOGRAPHY

Anamosa. COURTESY BECKY DIRKS HAUGSTED

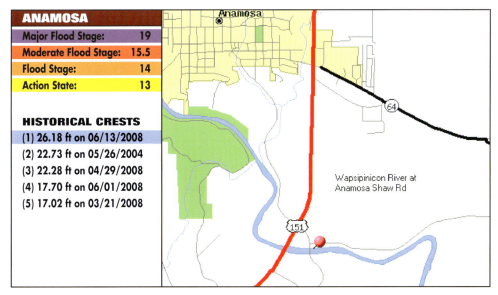

COURTESY NATIONAL WEATHER SERVICE

WHISKED AWAY BY THE WAPSI
Anamosa

When news of the impending flood reached Anamosa, city officials scheduled daily planning sessions at the Jones County Courthouse. The anticipation was the floodwaters would equal flooding seen in 1999. But unexpected forces of nature threw a monkey wrench into their expectations. Anamosa city administrator Patrick Callahan recalled four factors that occurred simultaneously to put the Wapsipinicon River into the history books. "We had six inches of rain that fell on Buffalo Creek beginning June 11 to June 12. The rain occurred just upstream of where it goes north and west of Anamosa," said Callahan. "Number two, the Wapsi was cresting about the same time and was full of water. The third factor was that all of the ground around the river was saturated and the water had no place to go. Finally, on the day of the sandbagging, which was June 12, we had an additional three inches of rain."

Anamosa. COURTESY DARYL SHEPANSKI Anamosa. COURTESY BECKY DIRKS HAUGSTED Decorah. COURTESY LUTHER COLLEGE

A call was put out for help. In the early morning hours of June 12, volunteer firefighters, city employees, and hundreds of residents came forward to try and save their city. There were even inmates from the state penitentiary in Anamosa on hand to help sandbag, while inside the prison more inmates worked filling sandbags.

Late in the afternoon, the situation took a turn for the worse. An elderly gentleman watching the river reported to Callahan that he could literally see the water rising. "I thought, oh no, this is bigger than what I was imagining," said Callahan. Indeed, the water was coming up so fast it was frustrating the efforts of the 200 or more volunteers who were sandbagging the town's levee. As they watched, their hard work was literally washed away. "The floodwaters had such velocity and current that as quickly as we put sandbags on the levee, the water was knocking them off. One of our levees then broke and it became obvious there was nothing we could do to plug the hole, and I knew the whole area was going to be flooded," said Callahan.

And it was. The city's wastewater treatment facility flooded, forcing a total shutdown and incurring approximately $3 million in damage. Three public works buildings were flooded, along with the city's recreational areas, which included the high school softball, baseball, track, and football fields. Three private businesses, two apartment buildings, and ten homes also had significant flood damage.

The town's concrete levee was built to withstand a 100-year flood crest of sixteen feet. When the water crested on June 12, it reached 18.5 feet.

The town never lost its water supply, thanks to the forward thinking city officials who had placed the water plant up on a hill. The next bit of good news came when an expected surge of water on Friday, June 13, never materialized. "The folks from Stone City, which is upstream of Anamosa, called and told us they had seen a second surge of water even higher than the one they had already experienced," said Callahan. "We thought maybe we would see another two to four feet, but apparently we were downstream just far enough that the water spread out and dissipated."

NORWEGIAN GULLY WASHER

Decorah

Decorah is the seat of Allamakee County in northeastern Iowa, home to Luther College. Along the north side of town runs the Upper Iowa River. The river, with its limestone bluffs, is a huge draw for canoeists, anglers, and hikers.

When the flooding of 2008 hit Decorah, city leaders decided not to take any chances. One day ahead of the crest, close to 450 house-

Sutliff Bridge, before. COURTESY JOHN NOST

Sutliff Bridge, after. COURTESY JOHN DEETH, *IOWA INDEPENDENT*

holds were told to evacuate, including three nursing homes. "That was quite a big challenge," said Jerry Freund, city administrator. "We evacuated at 12:30 at night. Several local churches volunteered space, as well as a middle school; a number of residents went to dormitories at Luther College. We just didn't know how much higher the water was going to get."

When the Upper Iowa smashed its old record crest set in 1993 at 14.35 feet, it was dramatic but not life threatening. The river crested on June 9 at 17.9 feet and threatened to top the town's levee built in 1955 by the Army Corps of Engineers. "We had to shut down one of the main bridges," said Freund. But unlike many other towns, Decorah's levee "held up just fine." Freund said the biggest problem was the concentration of heavy rainfall. "The damage that was done in the community was almost entirely due to the rainfall and not the river itself. The ground was so saturated the basements were flooded," said Freund.

Much of the damage that did occur was to public infrastructure. Roads were washed out, and the town's wastewater treatment plant was damaged.

Luther College also saw flooding on campus. The liberal arts college has a privately built levee that runs along the northwestern side of the campus. A portion of the levee was breached, and water stretched into sports fields and some buildings on the campus's north side.

BRIDGE OVER TROUBLED WATERS
Sutliff

One of the more charming tourist attractions in Cedar County is the Sutliff Bridge. Built 111 years ago approximately eight miles south of Lisbon, the bridge features three arching steel spans. When the raging Cedar River began pounding against the bridge, local residents stood by and fearfully watched the bridge struggle to remain upright. Among those was Doris McElmeel. On June 13, at just twenty minutes past the noon hour, McElmeel stood with her camera ready and watched as one of the bridge's arches lost the battle. "What was so amazing was that it went so fast. It crackled and just collapsed and went peacefully in the water. It was unbelievable that something that big could be destroyed in a matter of minutes. I just had time to click four pictures and it was in the water," she said. While the other two spans remained, McElmeer said it was still "a very sad historic loss for the community."

RECORDS ARE MADE TO BE BROKEN
Shell Rock

Perhaps the oldest record in Iowa to be demolished by the Flood of 2008 occurred along the Shell Rock River, near the town of Shell Rock. There, on June 10, the water rose 2.3 feet above the benchmark crest established during the pioneer-era flood of April 1, 1856. ✸

CREDITS LEFT TO RIGHT: KATHY KAUFMAN, IOWA DEPARTMENT OF TRANSPORTATION, JOEL ABRAMS, JOSEPH L. MURPHY—IOWA FARM BUREAU, JOSEPH L. MURPHY—IOWA FARM BUREAU

THE END of the LINE

ROAD AND AGRICULTURAL IMPACTS

From Iowa roadways to farm fields, the historic floods of 2008 wreaked havoc and destruction. According to a survey by the Iowa Department of Agriculture and Land Stewardship, there was an estimated $40 million dollars in damage to conservation practices from record rainfall and flooding. That shakes out to about 2.3 million acres, or 10 percent, of Iowa's cropland suffering severe soil erosion. "It will take years to rebuild what was damaged," said Iowa secretary of agriculture Bill Northey.

WHAT ARE FLASH FLOODS?

A flash flood is a rapid rise of water along a stream or low-lying urban area. Flash flood damage and most fatalities tend to occur in areas immediately adjacent to a stream or arroyo, due to a combination of heavy rain, dam break, levee failure, rapid snowmelt, and ice jams. Additionally, heavy rain falling on steep terrain can weaken soil and cause debris flow, damaging homes, roads, and property.

Flash floods can be produced when slow-moving or multiple thunderstorms occur over the same area. When storms move faster, flash flooding is less likely since the rain is distributed over a broader area.

WHAT ARE RIVER FLOODS?

A flood is the inundation of a normally dry area caused by an increased water level in an established watercourse. River flooding is often caused by:
* Excessive rain from tropical systems making landfall.
* Persistent thunderstorms over the same geographical area for extended periods of time.
* Combined rainfall and snowmelt.
* Ice jam.

FLOOD GRAPHICS ABOVE COURTESY NOAA DEPARTMENT OF COMMERCE

The cost to Iowa's roads is at $30 million and still accelerating. In all, flooding forced the closure of some 464 miles of the state's highways, including 303 bridges and culverts. County roads were also impacted and, according to Dena Gray-Fisher, media marketing manager of the Iowa Department of Transportation (DOT), there will be additional associated costs.

River ravaged road—Highway 1 south of Mt. Vernon.
COURTESY IOWA DEPARTMENT OF TRANSPORTATION

Road closures peaked on June 13. On that day, said Gray-Fisher, fifty-one state highways were closed, primarily in eastern Iowa. "To put that in perspective," said Gray-Fisher, "we issued 153 new press releases about road closures. We set up an emergency phone bank to take calls from travelers, and in a nine-day period logged 12,000 phone calls."

An Iowa DOT website that showed flood pictures and images from real-time cameras also logged hundreds of thousands of Internet visits during that time. In addition to their regular duties, Gray-Fisher said DOT personnel hauled pumps and barricades, filled sandbags, assisted in traffic control, and helped provide critical items to communities. For example, Alliant Energy needed to bring in transformers and a boiler to Cedar Rapids. The DOT assisted in finding a route for the oversized loads and escorted the trucks. At the same time, a moratorium on oversized loads coming into Iowa was issued, so truckers were forced to find a detour around the entire state. "What we learned the most," said Gray-Fisher, "was how important . . . and it reminds every Iowan . . . how important the infrastructure is for our economy, our personal business, and our everyday lives."

APPLES AND ORANGES

100-Year Flood vs. 500-Year Flood?

According to the U.S. Geological Survey, the 100- and 500-year flood "is a calculated level of floodwater expected to be equaled or exceeded every 100 or 500 years on average." In other words, there is a 1 percent chance (1 in 100) in any given year of experiencing a 100-year flood. Comparably, there is a 0.2 percent chance (1 in 500) in any given year of experiencing a 500-year flood.

It should be stressed that the 100-year and 500-year floods are independent events from the perspective of probability. That means if one of those events occurs, it has no effect on future events occurring. In other words, if a 100-year flood occurs, that does *not* mean that people are "safe" for the next ninety-nine years. The risk of experiencing a 100-year flood in any given year is the same, regardless of whether or not it occurred recently. So, while it is highly unlikely, it is remotely possible that a 100- or even a 500-year flood could occur in successive years.

HOW HIGH'S THE RIVER, MAMA? MEASURING THE RIVER

When the Cedar River crested in Cedar Rapids at thirty-one feet, did it mean the river was thirty-one feet deep? Not necessarily. The reason why is that river stages are measured in relation to elevation. Since elevation is not the same from location to location, a consistent starting point allows stages to be measured in a way that's easier for people to conceptualize, understand, and use. This process is known as the zero datum method.

To employ the method, a reference point or stage is selected at a known elevation relative to sea level. This is the "zero datum level," and this point is usually established slightly below the riverbed. Any river measurement above this point is a positive number. Anything below would be a negative number associated with low water levels. Since the datum level doesn't typically correlate with the bottom of the river, a measured stage of thirty-one feet could reflect a river depth that was somewhat lower. The bottom line is that by using the zero datum approach, we can keep true sea level elevations out of measurements, allowing river stages to be recorded in much smaller numbers.

Here is an example of the zero datum method. In Cedar Rapids, the stage level elevation is 706.9 feet above sea level. Instead of reporting the river stage as 706.9 feet mean sea level (MSL), a zero datum level was set at 700.5 feet MSL. The difference in elevation above the zero datum is 6.4 feet, which is the reported river stage. (Most people agree a figure of 6.4 feet is an easier figure to manage and visualize than 706.9 feet.) Additionally, if you know what flood stage is (twelve feet in Cedar Rapids), it's basic math to see the river has to rise 5.6 feet to reach flood stage, and 24.7 feet to reach the 2008 crest of 31.1 feet. ✺

COURTESY NOAA DEPARTMENT OF COMMERCE

CREDITS LEFT TO RIGHT: TRAVIS MUELLER, MERCY HOSPITAL CEDAR RAPIDS, TRAVIS MUELLER, DOUG AND LISA BOCKENSTEDT, KWWL SHOW AND TELL 7

FLOOD STORIES

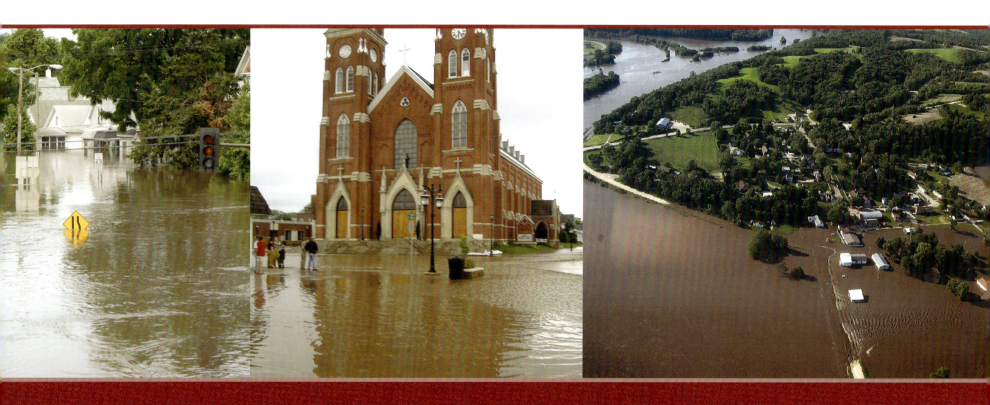

The smell of a river turned into a toxic cocktail is something they will always remember. The slippery mud was worse than walking on ice. But what will endure after homes are rebuilt and lives put back in order are the stories. For the tens of thousands of people devastated by the extraordinary Flood of 2008, the river will not be remembered so much for its cruelty but for the kindnesses it inspired. Because, in this disaster, the river didn't divide, it brought together all that is good: courage, resiliency, the generosity of strangers, and the love of families.

These are some of the survival stories that offer hope and inspiration, stories worth reading and re-reading.

Cedar Rapids. COURTESY TRAVIS MUELLER   Coralville. COURTESY CITY OF CORALVILLE   Shrock residence, Iowa City. COURTESY TIM HANSEN

A STITCH IN TIME

When it was inevitable that floodwaters would lap against his business, Dave Axline became creative in his flood protection efforts. "I duct-taped and caulked the door seams to prevent the water from coming in," he said. At the time it seemed like an excellent idea. West Side Sewing is a third-generation (just on the edge of going to a fourth generation) family business. Axline's daughter, Shelley Cervantes, is poised to take over. For eighty-one years, the store had sat within one block of the Cedar River and never had a drop of water inside.

If crest projections stayed around the twenty-one to twenty-two-foot level, Axline's flood prevention methods would have succeeded. But nothing, not even duct tape and caulk, could have kept the Cedar River out of West Side Sewing when the crest hit 31.1 feet. "All three of our buildings had ten feet of water inside. Everything inside was gone, destroyed and damaged beyond repair. The water was within six inches of totally covering the West Side sewing buildings," said Axline. Only 20 to 25 percent of the inventory was saved.

Among the lost items were huge 200- to 300-pound sewing machine cabinets with brand-new machines inside. The strength of the current flushed approximately twenty-two of the cabinets and machines out the broken windows and downstream. Even items stacked eight feet high on shelves were submerged.

When the river levels dropped and Axline and his employees were able to return, they were amazed by what they found. "The river had done things you could never imagine," said Axline. "Bottles of cleaning supplies at the Oreck Center were hanging from the ceiling furnace vents. All 7,000 square feet of ceiling tile was a mucky paper mâché mess on the floors, covering the fixtures, products, carpeting, and office equipment. The comment we heard most was, 'I cannot believe what I see.'" After working all day to clean the vile mess, Axline was grateful for the simple pleasure of being able to return to his dry, comfortable house. "As I headed home through the residential area, I saw a sight that put things in perspective quickly. A young lady sitting on the curb in front of her home crying and folks trying to comfort her. People, hopelessly sorting, piling, carrying things from their homes . . . all with the same sad expression on their faces. Thank you God for letting me have a home to come back to. In the past, I could not imagine what it would be like to lose my home and possessions. Now I know a little of how they feel. No words, TV, newspaper articles, or other descriptions can relay that seemingly hopeless, empty feeling," he said.

WE LIVED, LOVED, AND LAUGHED THERE

In 1997, Greg and Linda Shrock retired to their dream neighborhood. The home at 821 Normandy Drive in Iowa City wasn't quite up to dream standards, but after the Shrocks put in a lot of old-fashioned elbow grease and modern-day renovations, it reflected its beautiful setting. "We adored that neighborhood. We lived, loved, and laughed and enjoyed life there," said Linda.

Fast-forward eleven years, and their beautiful street was transformed into an ugly, stinking, muddy mess. Flooding from the Iowa River invaded homes and streets, filling the Shrocks' house with four feet of polluted brown water. "This is the most emotional thing I have ever been through in my life, and I am a two-time cancer survivor," said Linda. "With cancer you know how to deal with it. This is completely different. You don't know what to do. There were so many things you have no control over. Sunflowers are growing out of my living room carpet. It's horrible and the smell is the worst thing I have ever smelled in my life. It smells like death, fish, and dirty mud. You feel like you can't breathe."

The Shrocks' neighborhood was evacuated Thursday, June 12, but with the help of their family, the couple had moved all their belongings out several days earlier. Greg was hospitalized the day before the mandatory evacuation, but despite all that happened the Shrocks consider themselves very fortunate. "We are among the lucky few. We did have flood insurance. We live seventy feet from the river and we weren't required to have it. But we are both in our mid-sixties and felt we needed it," Linda said.

Once the water had withdrawn enough that it was possible to wade through the neighborhood, Linda made the decision to go back to the house. A friend had urged her to open windows to reduce the growth of mold. "I understand the police didn't want us there, but I had to get those windows open, so I walked in the back way," she said. "Honestly, when I opened the door for the first time, I thought I was going to faint."

Linda had taken her cell phone and, with Greg just home from the hospital, she was keeping in close contact with him. After she finished the job, Linda called a few of her neighbors and offered to push their windows open, too. "They said 'yes' so I went into a few homes and all the furniture had washed to the center of the rooms. I couldn't do a whole lot to help, but I got the windows open. I was scared to death because if I fell or got hurt, I was on my own," she said.

Linda safely navigated out of the flooded neighborhood, not returning until the waters had completely receded. "It was heartbreaking," she said, referring to her neighbors who had lost everything. "Everything the water touched had to be thrown away. So their lives just lay there on the curb for the people of Iowa City to pick up and put in a truck," she said. "I will always have the utmost respect for flood victims."

Lights of a flooded city—Cedar Rapids. COURTESY ANDREW LAIRD

For weeks, the scene along Normandy Drive was like a battle zone, with mud and filth coating every surface. Workers on bobcat skid loaders hummed up and down the street. Inside homes, drywall was being ripped away, exposing the framework. Everywhere, volunteers wearing respirators, knee boots, and gloves carted away mold-infested insulation, doors, and carpeting. "There was no electricity and no toilets and no water," said Linda. "It was hot and nasty smelling. Emotions were running so high from everyone there. People were dragging all kinds of things to the curb. It could be your grandmother's rocker or a quilt that was special. It was all being hauled away."

The one bright spot, said the Shrocks, were the volunteers and the city crews. From the people who worked day after day hauling heavy furniture and soggy carpet to the curbs to the volunteers who brought in food, water, and helpful hands, "it was fabulous what they did," said Linda. "They were wonderful."

The Shrocks are not returning to their home on Normandy Drive. They hope to sell it, or perhaps rent it for a while. They are relocating to a home that sits high on a hill, where the only running water comes out of a faucet. "It is too emotional to go back there," said Linda. "We are not willing to take a chance that it won't happen again." Because the Shrocks do believe the waters will come again, in the not-too-distant future.

THE DAY OF OUR MISERY

He could appropriately be called the maverick of Czech Village. Greg ArBuckle, who owns the Tattoo and Piercing Emporium, is described by neighbors as a man who marches to the beat of his own drum. So when the floodwaters came calling, they weren't surprised ArBuckle chose to stay and ride it out. "We decided that the flood was not going to be that bad, they were talking on the TV and radio of only 21.5 to maybe 22 feet over flood stage. Well, in 1997 we didn't even have a wet basement after that flood," said ArBuckle.

"So we decided to set it out, that way if anyone came a-looting, I could read them from the double ought buckshot," he added.

When the water started rising, ArBuckle said it came in a rush for the first two feet, carrying with it a thin film of diesel oil. When the radio began reporting new levels as high as twenty-eight feet, ArBuckle was "stunned" by the news and turned off the radio.

Stocked with bottled water and canned goods, ArBuckle and his small family took shelter in the upstairs rooms. "I started taking pictures out of our windows of the other businesses on the avenue. The water just kept rising up. Every picture I took I said to my wife, 'It can't get any deeper than that.' Boy was I wrong! It was very strange to have no noise on the avenue, dead quiet!" he added.

For a long time he watched the water rise and a parade of objects float by, from small sheds to telephone poles, railroad ties, parts of roofs, and small uprooted trees. "There was even a body, I thought, but it was a mannequin," he added.

With no television or radio (he didn't want to wear the batteries

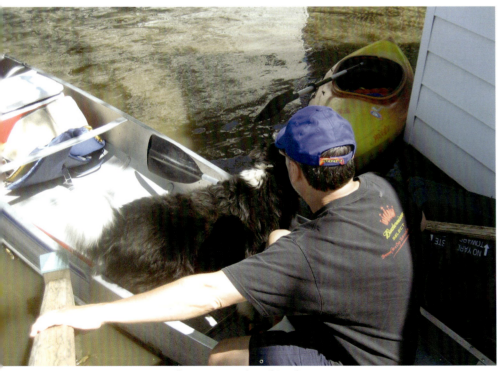

Saying goodbye to man's best friend. COURTESY CATHY MCROBERTS

Calm waters in Coralville. COURTESY CITY OF CORALVILLE

Fishy business in Waterloo. COURTESY RYAN CORWIN

Attention to detail. COURTESY BILL WITT

down, so only had it on the WMT 600 station for news), ArBuckle said he and his wife talked and watched people in boats go by. "Some were firemen, but others were sightseers out for the thrill of boating down the avenue. A large pontoon boat with thirty people having a beer party was out on the tour, too!" he added.

"We watched our Caravan go under water and saw it bobbin' around in the parking lot a little. I knew it was toast when the water went down." It was even more difficult, said ArBuckle, to watch the water destroy what so many people had worked so hard for their entire lives. "It was and still is heartbreaking," he said.

The couple had planned to stay in their home until the flood receded, but when the news reported the water could stay up for weeks, they capitulated and left. "So we flagged the next boat down and hitched a ride out of Czech Village; we were the last to leave. The firemen had gotten our neighbor earlier and she was as sad as we were. It really has changed my thinking about the river and how close one lives to it," he said.

After returning to their home, the ArBuckles found they had more to contend with than just mud and sludge. "We watched the vultures come around picking through the piles, taking what they want. You chase them away, they just come back. And the vultures in their $80,000-plus cars driving around looking and holding their noses and taking pictures, how crass they can be. Even had a family stop and have their children stand next to the sidewalk to have the devastation as a backdrop.

"I hope they enjoyed the day of our misery; we paid well for it. I think I will hit the next person that says to me, 'well you got your health anyway.' It would be best if they say nothing at all."

PACKING PICASSO

When the Iowa River grew threatening, there was no hesitation on the University of Iowa Arts Campus. Tuesday, June 10, the museum's insurance underwriters, Lloyd's of London, triggered the museum's disaster plan, rushing in a crew with fine arts expertise. Their job: to expedite the evacuation of the 13,000-piece collection that includes works from Pollack, Picasso, and Kandinsky.

Each piece had to be inventoried, assessed, and packed according to its needs. Acid-free wrapping, crates, and bins were secured. First

"Higher education"—moving books out of harms way. COURTESY UNIVERSITY OF IOWA

Waterproof fortress. COURTESY UNIVERSITY OF IOWA

Mustering the clean-up crew. COURTESY UNIVERSITY OF IOWA

to go was the museum's signature piece, the Jackson Pollock painting titled *Mural*. A crate specially built in Chicago for the painting was dispatched. It took eight strong men to lift the painting that measures nearly eight feet by twenty feet, and weighs approximately 500 pounds. They grappled with the unwieldy canvas that bowed and writhed before finally placing the painting face down in its special crating. Upright, the box was wedged through doors to the loading dock, a feat not easily accomplished. But finally the painting was safely stored in a temperature-controlled, specially cushioned truck and hustled off to an undisclosed location.

With the waters rising swiftly outside, a team of art professionals, UIMA staff, student employees, and museum volunteers moved even more quickly inside, wrapping, packing, and crating thousands of pieces of art in just two and a half days. "The staff worked tirelessly, some even spending the night," said Pamela White, Interim Director of the UI Museum of Art. The first of three semitrailer trucks left Wednesday, June 11, for a storage facility in Chicago. On Thursday, workers and volunteers hoped for two more full days in order to finish moving the art, but early Friday morning, at 7:45, the call came. They had to leave. It was a mandatory evacuation.

But the grueling work had paid off. In just three days, about fifty people had labeled, carefully packed, and shipped 10,000 pieces of art, representing approximately 80 percent of the museum's entire collection. The remaining 20 percent had been moved to the highest possible places inside the museum.

When the Iowa River crested on Sunday, June 15, the two buildings on the arts campus were heavily damaged. The Arts West Building, designed by Steven Holl, had filled with four feet of water on the first level and its basement was flooded. The museum also had a flooded basement, plus three and a half feet of water in its sculpture court and three to four inches of water in its three main galleries. The estimated cost of the damage to the two buildings was approximately $16 million.

When the waters receded, museum staff worked quickly to remove the remaining 20 percent of the collection. Fast-growing mold dictated speed and efficiency. "We had a quarter inch of mold growing every hour," said White. The floodwaters touched only a handful of pieces, and they were sent along with the rest of the remaining collection to the Chicago Conservation Center "to be triaged," said White. "When we think that we didn't lose anything, it is just amazing . . . amazing. We take our caretaking very seriously."

A CALL TO SERVICE

When the flood stampeded through Cedar Rapids, forcing evacuations, destroying buildings, and sweeping away possessions, there was also a surge of support flowing *into* the city. From Monday, June 14 through June 23, thirty-four different outside law enforcement agencies showed up to volunteer approximately 12,500 hours of service. Help came from Minnesota, Nebraska, and all across the state of Iowa. Eighty officers arrived from Minnesota alone. The U.S. Coast Guard, Iowa Air National Guard, Iowa Navy Reserve, Army National Guard, and the Iowa National Guard also arrived. At the peak of the flooding, there were 700 National Guard soldiers working to secure the perimeter of the flood zone. They brought their own vehicles, trailers, and supplies and were embraced by a grateful community. "Residents brought them hamburgers, hot dogs, baked goods, and coffee," said Sgt. Cristy Hamblin. "In the past we had never really helped other agencies outside of Linn County," she added. "That's changed now."

Police Chief Greg Graham had only been on the job for ten days when the flood forced the evacuation of the Cedar Rapids Police Department. The water pressure ripped doors off their hinges and filled up the basement. Riot gear, including helmets, padding, and coats, were swept away. The police department relocated to a nearby ice arena.

The floodwaters also pushed the dregs of society to the top. A handful of residents attempted to scam money from FEMA while others tried to register for a new driver's license that would reflect a flood-impacted neighborhood in an attempt to collect aid. But that didn't fly in this Iowa community with its solid Midwest values. "Residents were calling in left and right to report people trying to get FEMA money or a new driver's license," said Sergeant Hamblin. "Family members were turning in other family members. They weren't going to put up with it. The community was not going to stand for it. It totally brought out my faith in humankind."

Among those arrested was an overly enthusiastic young man from Marion, Iowa. On Friday, June 13, Sergeant Hamblin was escorting city council members through a flood-affected neighborhood when she met a National Guard soldier named Brandon Adams. "He approached and gave me information about the officers staffing the patrol boats," she said. "He seemed a little too gung ho. A little over the top." The young man told Sergeant Hamblin, "I am in charge." Along with his attitude, it appeared he was also taking advantage of his position by "making his subordinates do push-ups," remembered Hamblin. After investigating the officer's information, Hamblin found it to be false. That sent off warning bells, but not enough to trigger immediate action. The gig was up the next day when an Iowa National Guard member noticed the epaulets on Adams's uniform did not have the proper insignias. On Saturday, June 13, Sergeant Hamblin again encountered the fraudulent national

Home sweet home—Cedar Rapids. COURTESY TRAVIS MUELLER

guardsman. This time it was at the temporary location for the police department. Adams was soon charged for impersonating a public official, a misdemeanor punishable by up to one day in jail and a fine. However, Adams did manage to get his thirty seconds of fame impersonating a guardsman when his picture appeared in the local newspaper escorting a man out of the river. Of course, he was in full uniform.

A MIDNIGHT EXODUS

At midnight the decision was made. It was the first time in Mercy hospital's 105-year history that patients had to be whisked to safety. It was something the Sisters of Mercy, who built the hospital on 10th Street in Cedar Rapids in 1903, could never have imagined. It was also something present-day hospital administrators would never have dreamed possible. "We never expected to get flooded, and we were," said Karen Vander Sanden, a spokeswoman for the hospital.

And it never should have flooded. The hospital, a 370-bed regional medical facility, sits ten blocks east of the downtown and is out of the 500-year floodplain. But on June 12, water began to enter Lundy Pavilion, the newest addition, built in 2002. Worried that the flooding would cause the hospital generators to fail, the call was made to evacuate 183

Midnight mission. COURTESY MERCY HOSPITAL AND JIM SLOSIAREK—*CEDAR RAPIDS GAZETTE*

Helping hands. COURTESY MERCY HOSPITAL AND JIM SLOSIAREK—*CEDAR RAPIDS GAZETTE*

Boating through the hood. COURTESY TRAVIS MUELLER

patients. First to be removed were the critically ill and the infants in the Neonatal Intensive Care Unit. "Before we evacuated them, we worked with the state to find proper placements based on their needs and the location of their family," said Vander Sanden. "Patients were transported to hospitals all over the state. It was quite a process."

Meanwhile, outside the hospital, hundreds of volunteers came forward to help sandbag the exterior of Lundy Pavilion, an all-glass structure. "People were standing in waist-deep water, working. We feared the glass would break, so walls were built about five feet high on the inside and outside," she added. Even so, the building took on several inches of water. Vander Sanden noted that without the help of the volunteers, the damage would have been much worse.

DIVINE INTERVENTION

Natural disasters seem to have a way of finding Pastor Sheryl Eash Campbell. Pastor Campbell is the minister of the St. Paul's United Methodist Church in Waterloo. When the Cedar River started to rise on June 9, Pastor Campbell had a "feeling in my bones" when she was driving home and saw the water coming up on the 6th Street Bridge in downtown Waterloo. "I had to drive through a few inches of water, and I thought, 'In a day this is going to be something that they don't realize—and that is exactly what happened," she said.

The pastor's bones have some experience when it comes to floods. In 1997, Campbell was a minister at a Lutheran church in Grand Forks, North Dakota, when a 500-year flood forced 90 percent of the town's population to flee. Devastating fires followed, destroying many of the downtown buildings. "People were pounding on the door in the middle of the night saying 'You have to get out.' Hours later you couldn't leave—you had to be boated out," she recalled. That disaster showed her a side of America she had never seen before. "Having been through the whole sirens, rescue boats, the National Guard evacuation, and the 'oh no, now what next' stage, I became acquainted with volunteers in America, how

Czech Village in Cedar Rapids. COURTESY TRAVIS MUELLER

Polehna's Meat Market in Cedar Rapids. COURTESY MIKE AND BARB FERGUSON

they come to help you because they want to, and how incredible that is," she said.

When the flooding passed through Waterloo, just as Pastor Campbell suspected, she knew the volunteers would come, and they would come soon. "I started sending out e-mails to everywhere I could think to send them . . . to my friends back in Grand Forks, and said 'We could use help,' and before I knew it they started calling and rolling in," she said.

Campbell and her congregation led the way in welcoming volunteer church groups, by providing shelter, food, and even hot showers. With a New Life Center built ten years ago, the church had the experience and the space to throw open its door and create a "home away from home" for the volunteers.

POLEHNA'S: END OF AN ERA AND FIGURINE PIGS

There used to be a place like Polehna's Meat Market on every street corner, a full-service meat market and deli that specialized in hand-cut meats, a charming atmosphere, and friendly service. By the time the Flood of 2008 hit, Polehna's, located in the historic Czech Village since 1931, was a page from the past that still did a thriving business in the present.

Mike Ferguson had worked at Polehna's as a young man, and in 1991 he bought the business. "We had people from everywhere who would come, and we had our very loyal customers in the neighborhood, too," he said. "We actually had a genuine wood-burning smokehouse that looked like something out of the caveman times."

On Wednesday, June 11, flood warnings escalated. Ferguson took the unprecedented step of ordering a refrigeration truck to store his meats from the basement freezer. "We didn't really think it would get that high. We thought we were invincible," said Ferguson. "I also thought if it does get that high, boy, is Cedar Rapids going to be in trouble," he added.

Late that afternoon the electricity flickered off, and Ferguson went home with plans to return at 5:30 the next morning to meet the refrigeration truck and remove the frozen meat.

That morning, before sunrise, Ferguson returned to find a heartbreaking sight. In less than twelve hours, water had filled up his basement and was waist-deep inside the first floor of the shop. "It was actually a very surreal moment. It was so dark and quiet, and here I was wading to get inside my business and tripping over things in the water," said Ferguson.

Part of Polehna's exclusive charm was its figurine pig collection. Hundreds of pig statues were displayed in every nook and cranny, and the flood had turned them loose. When Ferguson pushed open the door to his store, plastic pigs bobbled and swayed, grinning at him in the filthy water. "At that point you think this is bad, and you know it's all over with. It would have been a lot more emotional for me, but I realized it's not just bad for me but for a lot of people," he said.

There was nothing Ferguson could do, so he collected his checkbook and the money in the cash register and left. The refrigerated truck was canceled. The meat in the freezers was left to the flood.

Two days later, Ferguson returned to Polehna's to begin clean-up. "In your mind, you are kind of mentally prepared for it," he said. "But it was a million times worse than you can ever imagine. It was like someone took your house and filled it three-quarters full of mud and water and then shook it like a snow globe and set it back down."

A customer favorite at Polehna's was an enormous glass jar full of fresh dill pickles that sat on the counter. Remarkably, Ferguson found it unscathed, sitting upright in the center of the market. "It was like, when all the flooding was done, someone had taken the jar and placed it perfectly in the middle of the floor. You could have taken a pickle out and eaten it right there," said Ferguson.

Joined by friends and volunteers, Ferguson scrubbed the meat market clean. The freezers full of spoiled meat were another issue. Ferguson had to line up a hazardous materials crew to remove it. It was a costly operation, but to Ferguson's surprise his friends arranged a fundraiser to offset the expense. Even so, the business was destroyed by the flood, and at age forty-eight, Ferguson does not want to start all over. "If I was twenty or thirty I would, but at my age and the idea of being $250,000 dollars in debt and working fifty to sixty hours a week, I don't have it in me," he said. "No one likes to be the person who closes down an institution in town, and unfortunately I have to be that guy."

RECLINERS ON THE RIVER

Smulekoff's Furniture Store sits one block from the Cedar River in downtown Cedar Rapids. It is considered the granddaddy of the home furnishings stores in the area. Started in 1889 by Henry Smulekoff, a Russian immigrant, the store carries a wide inventory of merchandise, including furniture, appliances, and flooring.

The building runs along 3rd Avenue and 1st Street, and features large showroom windows measuring approximately five feet wide and twelve

Take a load off! COURTESY CATHY McROBERTS

Construction delay. COURTESY DAVE WIELAND   Gnomes on edge. COURTESY CATHY McROBERTS   Problem on the porch. COURTESY TRAVIS MUELLER

feet tall. With only one corner of the building that actually sits in the 100-year floodplain, store employees thought the situation was manageable. "Our worst-case scenario was that no more than a half inch of water would reach the first floor," said Theresa Blair, advertising manager. To be on the safe side, employees moved as much furniture as they could from the basement level to upper floors. Pumps were installed inside the elevator shafts, and on the first floor, rugs were removed and stored on shelves. A few employees remained that night to babysit the pumps. Blair went home, confident that all bases had been covered. Sometime during the middle of the night, the unimaginable happened. The rapidly rising water shoved its way inside and forced the evacuation of the store's skeleton crew.

Thursday morning, Blair turned on the television news and was shocked. "You could see it was crazy. I was doing pretty good until my father-in-law called and had me flip news stations. Then I saw our furniture store, and you could see furniture coming out of the building and going downstream. I cried," said Blair.

Either the force of the water or the debris floating in the river had smashed the elegant storefront windows. A steady parade of chairs, couches, and tables, price tags attached, marched out and spun away. "It was all ruined. We wouldn't have wanted it back," said Blair. Water inside the first floor reached seven feet nine inches, destroying grandfather clocks, floor coverings, and entertainment centers.

There was a silver lining. Because the showroom windows had been broken, air could circulate throughout the building, minimizing the later impact from the floodwaters. "We did not have mold to speak off," said Blair. "We were able to clean up fairly quickly, and as for the smell, that is what amazed people the most. You could walk by some buildings and they reeked, but our building never did."

CAN YOU HEAR ME NOW?

Despite its location several miles south of Cedar Rapids, flooding from the Cedar River still impacted the Eastern Iowa Airport's communications system. The city-owned airport is tied to Cedar Rapids' communications hub located in the downtown (along with much of the city's other infrastructure). When flooding shut down the communications hub, the phone lines into the airport were also severed, making public contact difficult. "The phones were out for two or three days," said Pam Hinman, airport director. "Until one of the staff members was able to reach a telephone carrier and get one line into our administration building and one line into the information desk at the terminal, it was difficult for people trying to find our information." The problem existed for over a week.

PROUD TO SERVE

The water was literally rising before her eyes. Officer Lynda Ambrose knew she was in a race against time. It was Thursday, June 12, and her job was to evacuate homes on the north side of Iowa City. "I was so surprised at how fast the river came up in just the two to three hours I was there," she said. While most residents heeded the call to move, one elderly couple who had lived along the riverbank for years refused to budge. "They had lived there in the 1993 flood," said Ambrose. As the waves lapped higher and boats launched to take people and possessions out, the couple's son pleaded with his parents to leave. He finally succeeded. "Just as it was getting dusk, they were the last boatload of people to leave," said Ambrose.

In the Idyllwild neighborhood, the scramble was on to haul heirlooms and other precious belongings to safety. But the overflowing river was swallowing up streets and cutting off access. "People were driving across lawns, trying to get in and out with their things," said Ambrose. "The urgency really came when the spillway filled and the water started coming over. Then there was no control by anyone and no one knew what was going to happen."

After the floodwaters crested, Ambrose boarded a patrol boat to make a sweep of the flooded neighborhoods. She floated down Normandy Drive, a neighborhood where homes back up against the Iowa River and trees and shrubs provide a parklike setting. "All I could see were the tops of houses," said Ambrose. "There were pairs of ducks floating between the homes and fish flopping to the top of the water." Surrounded by green trees, birds, and ducks, Ambrose felt like she was floating through a surreal park. "Everything was greened up. The water wasn't that nasty. There were no oil spills when it first came up. It smelled like a river but not a rotten river."

The boat patrol's job was to prevent looting. But that was a guessing game. Some residents in the neighborhood had created a path marked with rope and reflectors to climb down a steep hill and check on their homes. "It was kind of an iffy thing what to do about the path. We didn't know if the people coming down were residents or looters," she said.

What resonates the most for Ambrose, apart from the massive destruction from the flooding, was the kindness of strangers. "I appreciate the people from the emergency services like the Red Cross and the Salvation Army who brought me a sandwich or a bottle of water while I was stationed doing traffic control," said Ambrose. Then there was the family of four, a mom, dad, and two little girls pulling a wagon filled with cookies. "They were handing out M & M monster cookies," said Ambrose. "The family baked them themselves and they were great."

Sandbags to the rescue. COURTESY CITY OF CORALVILLE

EXPEDITIOUS ABOUT SELF-PRESERVATION

When Gloria Gellman bought her condominium in 2000 in the Idyllwild subdivision in Iowa City, there was never a hint of danger from the nearby Iowa River. "I had no qualms at all," said the eighty-one-year-old grandmother of three and great-grandmother of one. The condo fit Gellman's needs perfectly. She was within easy distance of her grown children, was less than a mile from the city center, and lived in a beautiful setting.

Then the flood came. Gellman, who lost her husband in January 2008, is an independent sort who does not procrastinate. So when the first sign of trouble threatened from the Iowa River, she didn't hesitate. "I am expeditious about self-preservation," Gellman said. "The first thing I did—and this is good advice—the first thing was to grab all my photograph albums, financial records, and other documents, and put them in the trunk of my car. The worst thing that can happen is to have your family heritage go floating down the river."

So while others waited and watched the river, on Tuesday, June 10, Gellman called a moving company, ordered a truck, and began the overwhelming job of packing up her household in a matter of hours. "My family was helping. They were all working in different rooms, along with the movers. Things were just stuffed into plastic bins. Nothing was even categorized."

She moved just in time. On Thursday, June 12, police officers evacuated the condominium subdivision. "I could see it lapping against the building, and then all of sudden it was there and we had to get out," she said. Efforts to sandbag the area had been fruitless. Gellman's first-floor condo was awash in three and a half feet of water. Her beautiful hardwood floor buckled to such a degree that it took on a bizarre, warped shape. "The FEMA man who came to look at my place said, 'Why do you have a boat in your living room?'" she said later. "My son had to saw the floor apart so we could get items out of the bottom cupboards."

Gellman considers herself to be lucky. She was able to save her family treasures, while many of her neighbors lost everything.

"I'VE GOT A VILLAGE"

The 100-year anniversary for the Saddle and Leather Shop in the Czech Village in Cedar Rapids was 2008. It was also the year the facility was forced to shut its doors. Massive flooding from the Cedar River destroyed the building on Thursday, June 12. "We had lifted everything up and stacked it high up on shelves," said Nan Barta, who owns and operates the shop with her sister Kristine Jones. "Wednesday night [June 10th] I drove in front of my building and went into the basement. Water had started to back up. Hours later, at 1:30 in the morning, the burgler alarm went off, and we had twelve feet of water in the building. I was in shock," Barta said.

A soggy sanctuary—Parkview Church, Iowa City. COURTESY PASTOR SCOTT STERNER

A farmer's plight. COURTESY JOSEPH L. MURPHY–IOWA FARM BUREAU

Raising water instead of corn (near Columbus Junction).
COURTESY JOSEPH L. MURPHY–IOWA FARM BUREAU

The current from the river was so strong it blew out the limestone foundation of the century-old building. "It would cost more than 50 percent of the assessed value to reclaim it, so we are going to have to demolish the building," said Barta. With its hitching rail, embedded horseshoes in the front sidewalk, new roof and walls, the building was supposed to last a lifetime—until "the river came and took it away." Although the building was destroyed, the sisters plan to continue the business.

Losing the building is like ripping out the middle section of a book on the Barta family history. In 1946, their father, George Barta, bought the business and worked for fifty years doing saddle and leather repairs. Even at ninety years old, George Barta would report for work. Both girls grew up in the store, learning the trade.

Inside, George Barta's antique horse memorabilia was displayed against the historical backdrop, where tools still hung on their original hooks from 100 years ago. The collection included more than 1,000 items, including sleigh bells, antique bridles and lariats, horse books, and a forty-two-inch-high paper mâché pony purchased in Germany in the 1870s named Sugar. Sugar survived the flood, but much of the collection and merchandise was damaged.

After the waters receded, a sign appeared in the window of the destroyed Saddle and Leather Shop: "Horse folks are the best." Without being asked, hundreds of volunteers from the local horse community had showed up with gloves, boots, and old clothes, ready to muck out and clean up the shop. "In four days' time, they had stripped everything out of the building and were working tirelessly to recover and reclaim our products and the antique horse collection," said Barta. "When they say it takes a village . . . well . . ." Barta paused a moment "I've got a village."

GREAT-GRANDMA'S GIFT OF LOVE

A great-grandmother from Cedar Falls, Iowa, Arlene Kressley is cherished by the children in her life. When the floodwaters trespassed into her neighborhood, Kressley was not worried. She had lived in her home for thirty-four years and never experienced a single drop of water from flooding. But on Monday, June 9, Kressley's grandson Ryan Shatek, an engineer with the city of Waterloo, called and urged his grandmother to move her belongings as high as she could and head out that night. "He insisted that I get out. He worried that if we got the water he expected, I would lose my vehicle," said Kressley.

Heeding his advice, Kressley left at ten o'clock that night. It was perfect timing. Soon after, flooded streets cut off access to main roads in her neighborhood and water invaded Kressley's house up to the first floor. "When it was all over, I could see the high-water marks on my house. But I was very fortunate and blessed. I didn't lose my house. Many of my neighbors did," she said.

When two of Kressley's great-grandchildren learned of her misfortune, they wanted to help because, as ten-year-old Chandler Adam said, she "means a lot to me." So Chandler and six-year-old Isaiah Robins set up a stand with cookies and pencils and asked for donations. With every gift, the donor received "two cookies and a pencil." The two boys raised $18.64 and planned to give the entire amount to Kressley. But when they learned a blind neighbor with diabetes had also suffered losses in the flood, they asked Great-grandma if she would mind if they split the money and gave half to the neighbor and half to her. "Their hearts were in the right place, and it really touched me to think they would do that for me and also help the neighbor," said Kressley.

Kressley's home was deemed unlivable. While her family repairs her house, she is staying with her daughter. "My son-in-law is driving 200 miles to come and help work on the house," she said. As for despairing of her situation, Kressley refuses to shed a tear. She is surrounded by a loving family, and besides, she said, "Crying would have just made more water."

A "CZECH"ERD PAST

Founded in 1974, the National Czech and Slovak Museum and Library is the nation's foremost institution on Czech history, art, and culture. Seated along the Cedar River in the heart of the Czech Village in Cedar Rapids, the center stretches into five buildings, including a restored 1880 immigrant's home.

On Tuesday, June 10, with the waters rising, the staff built sandbag levees several feet higher than the expected crest. The following day they added more precautions. Two semi loads of artifacts were moved to higher ground and other materials were stored in the attics of both the museum and the collection center. A short time later, the staff was evacuated. Even then no one was overly concerned, thinking the mandatory evacuation was just a safety precaution. "In our minds, we really thought we had secured the facility and there would only be inches of water. In fact, we thought we had taken one or two steps further than what would be necessary," said Leah Wilson, director of marketing and communications.

It wasn't until the river crested on the 13th and the museum campus was visible on aerial photographs that the full impact of what was happening sank in. "You could clearly

National Czech and Slovak Museum, Cedar Rapids. COURTESY TRAVIS MUELLER

Cedar Rapids under siege. COURTESY DARYL SHEPANSKI

When Wilson walked up to the museum entrance following the flood, she was horror-struck to see that a wooden kiosk had been used like a battering ram to smash through a front window. On the other side hung the chandelier. "Looking through the shattered window, I could see the crystal chandelier—and it was perfectly intact," said Wilson. "A lot of people asked about it, and it was really heartening to see it had survived."

Other small pleasures inspired the staff throughout the recovery process. The immigrant home, despite sitting in water for two days, was intact. "They really knew how to build houses back then," said Wilson. "It was in remarkably good shape."

A painted ostrich egg for sale in the museum store was also a remarkable find. "It was sitting at my feet, absolutely unharmed, in the museum door. It had gotten all the way here through heavy debris in the water. Next to it was a medium-sized crystal vase, absolutely intact," said Jan Stoffer, director of operations and education.

see the water was above the front entrance doors and you could barely make out the top of the immigrant home. We knew then we were dealing with a major catastrophe," said Wilson.

Estimates put the floodwater as high as seven feet inside the buildings and ten feet on the outside. Because of its proximity to the river, a swift current, so forceful it kept boats away, ran through the campus.

When the staff was finally allowed back in, Wilson was overcome by the noxious odor, which was overpowering from blocks away. "I will never forget that smell. It wasn't like anything in nature. It was a toxic cocktail. It would sting people's eyes and you could feel your skin itching and burning," she said.

One of the crème de la crème pieces of the museum is a 410-pound chandelier composed of 600 bohemian crystals. It was constructed on site, a favorite among tourists, and a powerful symbol of the museum.

Museum exhibit cases were also discovered in interesting positions. According to Stoffer, the wooden bases floated out from beneath the plexiglass cases, upending them. River water then filled up the plexiglass, transforming the cases into small floating aquariums. Little was damaged, however, because the water inside cushioned the artifacts, preventing breakage. "It's funny how the water works," said Stoffer.

While 80 percent of the NCSML's collection was preserved, approximately 20 percent was impacted by the flood. Textiles were sent to Chicago for preservation and conservation efforts, including a folk costume called a *kroje* that is embellished with fish scales.

Even when all signs of the historic Flood of 2008 have been erased, the National Czech and Slovak Museum and Library intends to preserve the memories for "the citizens of Cedar Rapids and the world." Stories and photographs that capture the spirit and essence of the flood are being collected for future exhibits.

Holy water. COURTESY TRAVIS MUELLER

HISTORY IN THE MAKING

When floodwaters peaked through the streets of Cedar Rapids, very few people had the opportunity to see firsthand the flooded neighborhoods. Mark Stoffer Hunter was among those afforded the unique opportunity to boat through impacted areas. Hunter is a historian who was hired by the city of Cedar Rapids to document the flooding. He was granted special access to areas otherwise closed to the public. During his water tour, Hunter said, he saw many bizarre things drift down the river. "The power of the water was so great flowing through a Clark gas station that it lifted three gigantic underground gas storage tanks and carried them downstream. One went as far as twelve blocks," said Hunter.

Norwood Promotional Products manufactures writing instruments and assorted other products. When the floodwaters erupted, ten feet of water flowed inside the plant, causing extensive damage. That day on the river, Hunter saw another byproduct of the Norwood flooding. "Ten blocks to the south of that company I saw literally tens of thousands of souvenir pens floating along," said Hunter. The cases of ballpoint pens had burst open and made their way from the Time Check neighborhood beneath Interstate 380 into the southwest section of Cedar Rapids and the Kingston area.

One of the most visual sights was a crush of small houseboats wedged against a collapsed railroad bridge. The houseboats had been tied up in a small harbor along the riverbank adjacent to Ellis Park, and about twenty-five boats had broken free when water surged over the levee in the Time Check neighborhood. "All those houseboats bobbing up against the bridge really added to the concerns of a possible collapse," said Hunter. However, the double-track bridge, built in 1898 with limestone blocks, did not buckle. A single-track railroad bridge owned by CRANDIC Railway and built in 1903 with concrete piers did fall victim to the rushing water. That bridge was located south of the dam.

What also struck Hunter was the quiet. There were no sounds of traffic or trains. Even the freeway was quiet. "There was just the sound of water," he said. Hunter compared the sight to an "eerie mix of ghost town and Venice, Italy."

As he boated in the late evening, the sky turned a deep orange and shadows began to play upon the water. Hunter was struck by what was happening in that moment of time. "As a historian, there is a sense that you are truly recognizing history happening right in front of you, history that would be talked about for generations. It was like seeing a battle or a major historical event."

TO THE RESCUE

The call came at approximately 2:10 A.M. on June 13: a man had fallen into Ralston Creek face down and was being propelled swiftly downstream in the high water. Heavy rains and backup from the flooded Iowa River had transformed mild-mannered Ralston Creek into a raging torrent. Iowa City police officer Matthew Huber responded to the call. The man had gone in near the New Pioneer Coop at 22 South. Van Buren Street. Huber moved quickly. The victim had disappeared, pulled under the water at the intersection of Washington and Van Buren streets. He resurfaced near College Street. Searching from a bridge, Huber couldn't believe his luck. "I could see the back of the guy's head and the back of his T-shirt . . . black in the brown water. He had a lot of things working

Road rage. COURTESY MARISSA REMETCH

Bridging the flood. COURTESY BILL WITT

High tide in Vinton. COURTESY DAN ADAMS

for him," said Huber. "All I could think about was getting down there as fast as I could. I couldn't tell what the terrain was like and just dropped down into brush and worked my way toward the creek."

When he reached the victim, another man, Sean Hansen, who had witnessed the man falling in the creek, was also there. "Sean had stopped the guy from floating down but couldn't bring him in," said Huber. "When I got there I picked him up so his face wasn't in the water and he could breathe." The two men and the victim were in about three and a half feet of water. "We had this significant downpour. Usually you can see the bottom of the river, but not this time. There was just a massive amount of water. It was the highest I have ever seen and the current was so strong. I couldn't see anything. You couldn't tell where you were going. The terrain was constantly shifting. After it was over, my shoes were full of rocks," said Huber.

From where the two men stood in the river, it was impossible to pull the victim to safety. The riverbank in front of them was blocked by an eight- to ten-foot-high concrete barrier. So they walked downstream, carefully keeping the man's face out of the water. "We finally were able to pull him over to shore and there was an ambulance waiting," said Huber.

"I could tell he was breathing but I don't know how he survived. He had been face down in the water for a pretty long time."

Paramedics handed down a backboard, and the victim was secured and carried to the ambulance. "It feels like everything took so long, but in reality it was minutes," said Huber. The victim was taken to University of Iowa Hospitals and Clinics, where he was treated and released. Officer Huber was awarded a lifesaving commendation. He was the first officer in Iowa City to ever be given the award.

DIARY OF DESTRUCTION

Pastor Scott Sterner is the pastor of adult ministries at the Parkview Evangelical Free Church located along the Iowa River in Iowa City. The following are excerpts from his blog (deo-gloria.blogspot.com) in which he chronicled the flooding that overtook the church on Friday, June 13.

Sat. June 7, 2008

We are not overly concerned about our facility flooding, but access to our facility is getting more challenged by the closing of Dubuque street and our parking could be easily cut by half if not more.

Next weekend we are planning on adding services with the hopes of distributing the population and hopefully reducing the parking problems that are sure to occur. The following are the times we'll be having services. In the mean time, pray that, Lord willing, we won't have much more rain. Depending on rain amounts we could surpass the flood levels experienced in 1993.

Sunday June 8, 2008

As services proceeded this morning, the flood waters continued to rise covering about 1/3 of our parking lot.

We'll keep you posted if the decision is made to begin sandbagging the church and regarding this weeks Vacation Bible School starting tomorrow morning at 9 am.

Later That Afternoon

From 3:00 to about 9:30 P.M. hundreds of people from Parkview, other churches, and our community showed up to help sandbag the church. It was a great picture seeing so many pitch in to help in a time of need. Unfortunately, we're only about half way in the sandbagging process. With the rain forecast, projections show us passing the flood of 93 levels so we're bracing for the long haul with this season of flooding.

Monday, June 9, 2008

Today we faced another long workday at the church. The most encouraging aspect of the day was the hundreds and hundreds of people who came together to help prepare the church for inevitable flooding.

We expect the emergency spillway at the Coralville reservoir to be breached within the hour (around midnight). At that point the floodwaters will no longer be controlled. Tonight's news broadcast called this the 500-year flood. We are bracing for the worst and will continue to raise our sand barrier tomorrow while also moving our offices to an alternative location.

Tuesday, June 10, 2008

Fortunately, levels did not rise as quickly as originally anticipated. Though flooding will increase in the days to come the Lord graciously granted us a few additional days to finish sandbagging and get our offices moved out of the church.

Wednesday, June 11, 2008

Today was another busy day. Parkview continued to function as a distribution center for our neighborhood throughout the day while my staff and I focused on moving out of our facility. The water began rising at a slightly accelerated rate throughout the day.

Today a diesel generator arrived that will power our pumping areas in the event of power outage.

For the second time in its history, the Coralville reservoir is now overflowing the emergency spillway. The water is now going to enter Iowa City at an uncontrolled rate of speed.

Thursday, June 12, 2008

The Lord gives and the Lord takes away. Today at 3:30 pm, in the midst of continued fervent efforts, city officials informed us that last nights heavy rains will result in the Iowa River cresting on Tuesday at 35 feet. This will put river water close to roof level. Hearing the news was difficult. All hopes on saving our facility were lost and emergency evacuation began immediately.

It is difficult to lose a facility that was laced with so many meaningful memories over the last 9 years, but in many ways the loss generates a deeper appreciation for those things that matter most, namely our faith in the sovereign hand of God and our appreciation for the body of Christ.

Later That Day

Our Boston grand piano was removed via fork lift over the flood waters.

A little after 9:00 pm, Jim Douglass (our administrator) and I shut down the water, turned off the master electric panel and walked out the northwest office door into ankle deep water. One chapter ends and a new one begins.

Friday, June 13, 2008

I want to start by thanking all of you for the kind comments, phone calls and emails. Your prayers have been an encouragement to everyone at the church. Many are still mourning the loss of a facility which hosted countless memories of weddings, funerals, services, and ministry events. Despite the loss, there is a sense of God's love and purpose that will drive us forward.

Church offices have now been moved to the Towncrest area in south-

PHOTOS THIS PAGE: Parkview Church—during and after. COURTESY PASTOR SCOTT STERNER

east Iowa City. In regard to services, we will be at West High School's auditorium at 9:00 and 10:30 am. Parking will be accessed through the east drive entrance. There is a chance we will run out of parking spots, which will require some to park in the large lot behind the school that is also being shared by National Guard vehicles forced to find new homes due to the flooding. Because of the parking situation, carpooling is recommended. I look forward to seeing many of you on Sunday!

Saturday, June 14, 2008

Today I had an extremely unique opportunity to get a look at the flood damage up close and personal. An individual in the church called with an offer for Pastor Jeff and I to join him in investigating the scene. As we were boarding a cameraman with CNN joined us. It is difficult seeing it up close and personal. We were very fortunate to have hundreds of people help evacuate most all of our equipment and resources.

Sunday, June 15, 2008

Today Parkview Church gathered at West High Auditorium for a very meaningful time of family worship. It was beautiful coming together as a community to worship our God and to experience such great unity grounded in our common purpose. There were tears, hugs, and laughter.

Four weeks ago those of us in the Worship Ministry thought it would be good to introduce the song "I Have a Shelter from the Storm." Here are the lyrics:

I have a shelter in the storm
When troubles pour upon me
Though fears are rising like a flood
My soul can rest securely
O Jesus, I will hide in You
My place of peace and solace
No trial is deeper than Your love
That comforts all my sorrows

Little did we know that the second week we would sing this song would be days after the tornado decimated the town of Parkersburg, Iowa. As is tradition when we learn a new song we planned to sing it again this weekend long before we knew of the 500 year flood about to hit our community and state. How amazing it was singing these words and realizing that our sovereign God is perfect in his timing. What a great source of truth and encouragement for such a time as this.

Look out below. COURTESY UNIVERSITY OF IOWA

Class canceled—Iowa Arts Building. COURTESY UNIVERSITY OF IOWA

A $231.75 MILLION PRICE TAG

In August 2007, the Iowa Board of Regents chose Sally Mason as its twentieth president. Ten months later, she led the university through one of its biggest challenges in its 161-year history. Despite flood protection to the 100-year flood level plus one foot, catastrophic flooding from the Iowa River swamped twenty UI buildings and caused an estimated $231.75 million in damage. Speaking at a press conference in Des Moines with Iowa governor Chet Culver, Mason said the university will overcome this adversity. "Rarely have our communities and our state been so tested, but working and standing together, we will get through this historic disaster," Mason said.

The $231.75 million price tag includes damage to buildings, building contents, tunnels, parking lots, debris removal, and leasing space.

Among the impacted buildings was Hancher Auditorium. Damage to the building and its contents totaled approximately $13 million. Inside the auditorium, floodwaters reached to row R. Flood marks could also be seen on the stage about a foot above the floor. But the new rallying cry for the Hancher staff evoked its spirit of recovery and rebuilding. For its 2008/2009 music series, venues were booked all over town and posters heralded the new philosophy of "Can't Contain Us!"

The Iowa Advanced Technology Laboratories also sustained heavy damages. The $25 million building was designed by California architect Frank O. Gehry and completed in 1992. Staff was able to move equipment, but floodwaters rose between two and three feet inside the first floor of the building.

University officials said the IATL, located on the river's east bank, sustained an estimated $8 million in building damage but $34 million in content damage because of its high-tech scientific labs.

The Engineering Research Facility was also in lockdown with water damage. The staff was temporarily relocated to the National Advanced Driving Simulator.

Other damaged buildings included Main Library, Power Plant, Iowa Memorial Union, Mayflower Residence Hall, and Danforth Chapel.

In the days leading up to the historic flooding, thousands of volunteers converged on the campus to help sandbag, load sand, and evacuate equipment. Young and old, Iowans and out-of-staters worked tirelessly to keep the river out. The Iowa National Guard also joined in the herculean effort. While the river still won, the effort put forth will be part of UI history. In a memo from the UI Flood Recovery Report dated June 19, Doug True writes: "Support from the University of Iowa community has been extraordinary. Students, faculty, staff, and townspeople alike have responded unbelievably. Recovery, however, will be a months-long process, years in some cases."

THE DAY THE MUSIC DIED

When Darren Ferreter was eight years old he saw an organ performance at the Paramount Theatre in Cedar Rapids. The "Mighty Wurlitzer" booming forth on stage transfixed the young boy. "I basically taught myself to play [the organ] because I was so intrigued by it," said Ferreter. The Wurlitzer built for the Paramount in 1927 was considered "the Cadillac of organ building." It was also one of fewer than forty organs still residing in its original home. As an adult, Ferreter played a pivotal role in the preservation of the Wurlitzer. He is a member of the Cedar Rapids Area Theatre Organ Society, a not-for-profit organization that promotes the continuing art of theater pipe organs.

When floodwaters swept into the Paramount, the Wurlitzer was submerged. There was eight feet and six inches of water up and over the stage, stopping only two feet from the bottom of the balcony seats. The pipes, which are stored in separate rooms, just escaped the flood. "The first time I went in to see it, it was an incredible shock," said Ferreter. "It had been concert ready. It was pristine and perfect, and to have it all suddenly pulled away was a shock." The 900-pound console was pulled off the stage with a crane and deposited in the Paramount's parking lot. But the stresses of being shoved on its back by the floodwaters were too much for the grand old lady. "It was the first time it had seen light in eighty years and it died right there in the parking lot," said Ferreter. Another factor speeding the Wurlitzer to her end was the fish glue used to put her together. Fish glue is considered preferable in building organs, but unfortunately dissolves in water, especially when soaked for days.

Over at the old Iowa Theatre, the 1928 Rhinestone Barton organ sat in water halfway up its console. It also was destroyed. The cost to repair both organs is $50,000 each. Ferreter said plans are under way to bring the music back. "They represent the public showpieces of the arts in Cedar Rapids," said Ferreter. "When the organs are back, the community will know the arts are back."

I HAVE NEVER SEEN SUCH POWER

For nearly two hours, photographer Travis Mueller literally floated through history. He was invited to join a rescue patrol going by boat through the most heavily impacted neighborhoods of Cedar Rapids. The day was Friday the 13th, and flooding was at its worst. "When we were out there the river actually crested," said Mueller.

Two firefighters, an official from the Department of Natural Resources, and Mueller staffed the sixteen-foot aluminum V-hull craft with a 30 horsepower engine. "It [the current] was about fourteen miles per hour going through the houses," said Mueller. "I have never seen so much power."

In the water below, things lurked that stretched the imagination. The boat plowed over small sheds, trucks, street signs, fences, and playground equipment. "You never knew if you were going to hit something and capsize . . . because the chances were, we would go over," he said. To add to the danger, not all the power grids had been shut off. Mueller

The Cadillac of organs—1927 Wurlitzer. COURTESY CEDAR RAPIDS AREA THEATRE ORGAN SOCIETY

and his crew were startled to chug up to a stoplight that had its red eye blinking on and off.

Boat patrols were still in the search and rescue phase of the operation, and Mueller had a first-row seat. "We went on a call of someone trapped in a building. Then we ran into people trying to rescue a cat out of a second-story window. After that, we saw two guys sneaking around in a canoe. It was pretty wild," he said.

In the midst of it, Mueller sought to capture the essence of the devastation on film. He photographed a historic church under water while a swimming pool floated by. Inside the swimming pool, it was dry. "It was unbelievable. The roles were reversed," he said. He photographed drowned houses, animal rescues, flooded restaurants, and the tops of street signs. "You couldn't believe it but by the same token you couldn't capture the emotion because it was so surreal," he said.

In his experience as a photographer, Mueller said he has taken pictures of air shows and "they are breathtaking." He has photographed hundreds of geese landing against beautiful backdrops, ice storms, wall clouds, and countless other awe-inspiring events. But the Flood of 2008 took him to a new level. "To actually capture nature at its utmost worst, I don't know how you call a flood like this magnificent but I have never taken pictures like that. It was unbelievable," he said.

The words of a DNR official struck Mueller by their simple truth. "Someone asked the official, 'What can we do?'" said Mueller. "He replied, 'We cannot contain this. We are just going to have to read about this in history.'"

POUNDING AT THE DOOR

In early June 2008, Elisha Shoars was looking forward to becoming a July bride. Plans were being finalized for a traditional ceremony, and the all-important wedding dress was hanging in her closet.

Then came the Flood of 2008. At one o'clock in the morning, June 10, Shoars and her finance, Chris O'Connell, were awakened by a pounding on their door. Floodwaters were swirling around their home on the north side of Cedar Falls. Told they had only ten minutes to leave, the couple gathered what they could and ran. The sandbag dike built behind their home had developed a small hole and water was gushing through. The couple, their children, and three dogs fled to O'Connell's mother's house in Waterloo.

The next day, they returned to try and retrieve more belongings. Using an inflatable boat, they rescued their black cat (named White) from the roof. Then they gave up. "The water was high and strong. The current almost took us away. So we decided that was enough," said Shoars.

When the water receded and Shoars came home, she found her beautiful wedding dress covered in mud and silt. "The part that tied around the neck was the only thing still white," she said. The dike behind the couple's home had been built with sandbags and clean black fill dirt. "We were the first ones to get the water and we got all that mud in there. It looked like fire and smoke damage mixed with a tornado," she said. Everything collected, bought, and stored away for their special day was ruined. The wedding was postponed.

That heartache was quickly followed by a worse nightmare. A snafu in record-keeping caused a mix-up with Shoars' social security information. According to Shoars, inspectors did not believe she owned her property, so she was not able to get much-needed assistance. "Oh my goodness, I am sure I have FEMA's phone number memorized," she said. "I have called well over fifty times, sent three different faxes, and spoken to their representatives in person." After two months of frustration, the problem was solved and she was cleared to receive financial assistance. "I was a wreck. I felt like the world was against us. I just focused on my family and God. Faith is a big part of getting through it."

The couple are in a temporary home, but a new wedding date has been set and this time the formal wedding is being replaced with a more casual event. "We are going to wear blue jeans and football jerseys and have a barbecue," said Shoars. "We are huge football fans. We always have a great time watching games. Plus, it is just too costly to replace all those items for the traditional wedding."

DISH TOWELS OF DISBELIEF

In disbelief, fifty-nine-year-old Thomas Sanders, a lifelong resident of Palo, wrapped a white dish towel around the handle of his front door. It was a beautiful spring day, and only hours earlier he had been sitting on his front stoop enjoying a cup of coffee. There had been no sign of creeping water from the nearby Cedar River, but the power had gone off and it seemed like it was time to leave.

As he drove through town, Sanders could see hundreds of white towels hanging from door handles, a signal to authorities that the homeowners had left. "This was Wednesday [June 11] and I thought I would be back the next day," said Sanders. He had lived through the floods of 1961 and 1993 and had never seen any water in his back yard.

That disbelief led Sanders to return a handful of hours later. He stepped out from his truck to ten inches of water in his driveway. "In just five hours, the water came up six or seven feet. I just couldn't comprehend how it came up that quick," he said. With the help of his nephew, Sanders quickly moved his valuables from the basement to upstairs. Items were stacked up high on beds, dressers, wherever there was an inch to spare. As he left for the second time that day, Sanders still had no idea what lay ahead of him.

On Sunday, June 15, Sanders joined hundreds of other anxious, restless residents at one of the main checkpoints outside town. National Guard trucks ringed the area, while Linn County sheriff's deputies worked to match names and residences with the growing crowd. "It was a lengthy process to be cleared to go in," said Sanders. "The emergency personnel had a checklist of houses that had been inspected and were structurally safe. If your name wasn't on the list, you couldn't go in," he said.

When he arrived at the checkpoint at one o'clock, Sanders' name was not on the list. The list was updated every half an hour and finally, at ten minutes past five, Sanders was allowed in. "I was told by the security people that curfew was seven o'clock. They said, 'if you're not out, we're coming after you.'" With only an hour and fifty minutes, Sanders quickly walked the mile into town. The towels fastened on doorknobs were all the colors of a dirty rainbow. "Some were black, others would be brown or tan and the upper portion white. You could see how much water got into the house by that towel," he said.

As he approached home, Sanders knew he had been dealt a harsh blow. "I could see the dirty water mark just below my picture window. I knew there had been considerable water in the house."

Inside, kitchen stools had floated down the back hallway. The refrigerator had been turned on its side. "The first thing I did was open every window and door. Then I grabbed some more clothes and I was basically out of time. I would like to have stayed. I had so much more work to do, but I had five minutes to make it back to the highway," he said.

The next day, the brutal scenario from Sunday played out again. Sanders arrived in the morning to find his name had been taken off the list. He waited another five hours before he was allowed back in. "That was frustrating," he said, "but you can't let it get to you. If you do, it has defeated you. I can't think like that. You persevere. You have things to do." ✷

Don't forget to stop! COURTESY TRAVIS MUELLER

UNPRECEDENTED RECORDS
Epic in Scope, Beyond Imagination

When historians attempt to get their hands around the Flood of 2008, they are going to find it slippery and difficult to grasp. So immense and all-encompassing were the floods in Iowa, that all-time record crests were established on nearly every river between Des Moines and the Quad Cities. Punctuating the feat is one enormous fact: not only were records broken, they were in many cases shattered. The rushing water was a zephyr so great it roared through the legendary benchmark crests of 1993 by as much as eleven feet.

Despite toppling 1993's record levels, however, there are some comparisons that can be drawn between the two greatest floods in modern-day Iowa history. According to Harry Hillaker, Iowa's state climatologist, you can find similarities by breaking down sections of the two floods. For example, in both 1993 and 2008, unusually wet weather preceded the floods by several months. "Both times there were long, cold, wet winters, followed by extremely rainy springs," said Hillaker. "The odd thing was that we had completely different weather patterns. In 1993, we had an El Niño year, and in 2008 it was a La Niña that prevailed, yet overall each flooding event was almost identical," he said.

There were also significant differences. In 1993, the flood occurred over months, while in 2008 it lasted weeks. The rainfall was also more concentrated in 2008, with about nine inches of rain falling in a fifteen-day period. "That single amount is the wettest fifteen days Iowa has ever had," said Hillaker. The other factor separating the two was the scope of the flooding. In 1993, the floods covered a vast geographical segment of the Midwest. In 2008 it was much more localized, with the heart of the flooding centered on eastern Iowa.

Initially, the wet and rainy conditions that set the stage for the Flood of 2008 were nothing new for flood-seasoned veterans who had survived the Flood of '93. They didn't panic or make special preparations when the initial warnings began. Why should they? They believed they had experienced firsthand the very worst Mother Nature could dish out. With typical Midwestern fortitude they had dealt with it.

Then the skies opened and it rained and rained in the days leading

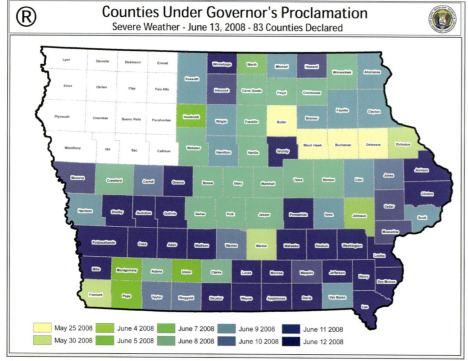

COURTESY IOWA HOMELAND SECURITY

TOP TEN NATURAL DISASTERS

EVENT	YEAR	FEMA FUNDING
Hurricane Katrina (AL, LA, MS)	2005	$7.2 BILLION*
Northridge Earthquake (CA)	1994	$6.961 BILLION
Hurricane Georges (AL, FL, LA, MS, PR, VI)	1998	$2.251 BILLION
Hurricane Ivan (AL, FL, GA, LA, MS, NC, NJ, NY, PA, TN, WVA)	2004	$1.947 BILLION**
Hurricane Andrew (FL, LA)	1992	$1.813 BILLION
Hurricane Charley (FL, SC)	2004	$1.559 BILLION**
Hurricane Frances (FL, GA, NC, NY, OH, PA, SC)	2004	$1.425 BILLION**
Hurricane Jeanne (DE, FL, PR, VI, VA)	2004	$1.407 BILLION**
Tropical Storm Allison (FL, LA, MS, PA, TX)	2001	$1.387 BILLION
Hurricane Hugo (NC, SC, PR, VI)	1989	1.307 BILLION

* Amount obligated from the President's Disaster Relief Fund for FEMA...as of March 31, 2006
** Amount obligated from the President's Disaster Relief Fund for FEMA...as of May 31, 2005

COURTESY FEMA

up to the record crests. Even with the constant downpours, residents clung to their memories from 1993 when the water never reached their back yards, never crept over the front step, and most assuredly never entered their homes.

Even as the water climbed higher and higher, they comforted themselves with scenes from 1993. Even if it became worse, it would only be by small degrees, not huge leaps. No one knew that in just hours those memories would come unraveled and be cast aside.

When the call came to evacuate, they collected a few things, turned off the lights, securely locked the front doors, and walked away, confident they would return the next day to an intact home. They didn't.

From seasoned forecasters to city leaders to residents, Mother Nature's power took everyone's breath away, throwing a curve ball no one could have anticipated. "This is the worst flooding I have ever seen. It is certainly historic and unprecedented," said Steve Kuhl, the meteorologist in charge for the National Weather Service in the Quad Cities.

Stunned people watched their homes, businesses, and, in one case, an entire town, nearly disappear beneath the water. The unimaginable became a fact. New levels of appreciation emerged. Those with flooded businesses were grateful to come home to a dry house. Those with flooded homes were grateful for friends and neighbors who sheltered them. Whole towns battling floodwaters were newly appreciative of support from law enforcement agencies, as well as volunteers, and church and service groups. Instead of sinking into a mud of despair, Iowans found new reasons and ways to bless their lucky stars.

"We have suffered a great wound," said Thomas Sanders, whose home was flooded in Palo. "But we are people of hardy stock. Not even the capricious Mother Nature can wring that from us."

In terms of economic and human impacts, the Great Flood of 2008 will indeed be uniquely defined as the most devastating flood in Iowa state history. Beyond that, the unprecedented back-to-back effects of the EF5 tornado and ensuing floods have elevated the storm-ravaged two-week period in Iowa to an elite national status. Following the tempest, eighty-three of Iowa's ninety-nine counties (84 percent) had been declared state disaster areas. Iowa governor Chet Culver indicated that damage to public infrastructure, such as buildings and schools, was still growing and nearing a billion dollars. According to damage estimates compiled by the Federal Emergency Management Agency, Culver also noted the billion-dollar figure could eventually rank Iowa's "un-natural" disasters as the tenth worst in U.S. history. ✸

EF5 tornado. COURTESY BLACK HAWK COUNTY EMA SPOTTER

500-year flood. COURTESY DAVID GREEDY, GETTY IMAGES

Iowa's Un-Natural Disasters of 2008

IMAGINE-DREAM-RISK-LAUGH-BELIEVE

Those five words are painted on the front steps of a home that I have passed by almost every day for years in Hudson, Iowa. Safe to say, those inspiring words embody the kind of hope, courage, and inspiration we saw Iowans exhibit in the days and weeks following the May 25 EF5 tornado and the record-setting floods of 2008.

The pain and misery suffered by so many will linger for years. While the financial losses are staggering, the personal agony and emotional toll have been nothing short of heartbreaking.

With tears in their eyes, the four Mulder brothers mourned the deaths of their beloved parents, Richard and Ethel Mae of Parkersburg. They were two of eight tornado deaths. The elder Mulders were inside the family's home in Parkersburg when the tornado hit. Unfortunately, due to physical limitations, Richard and Ethel Mae weren't able to get to their basement.

The storm took their lives and, with inhumane disregard, left their bodies strewn about their south-side neighborhood. The tornado didn't just take their lives and ruin the Mulders' home, it obliterated the entire structure. The winds, in excess of 200 miles an hour, even ripped up the sub-flooring and floor joists. Only the basement walls remained. That kind of tornado devastation is not often witnessed in Iowa. It's fairly common to see Iowa tornadoes take off roofs and knock down side walls. But when you see only basements remaining and bark stripped off trees, you know a rare, frightening, and bizarre display of power has just occurred.

Like the tornado, the record floods of 2008 left Iowans in a daze, too. So many times, I watched weather-weary storm victims survey their losses and search for the strength just to ponder their futures. "Where do we even begin," seemed to be a common question for hundreds of Iowans, many of whom lost everything, including personal items like irreplaceable photographs and other treasured memorabilia.

While these disastrous events could easily have symbolized our weakest hours, gritty, dedicated, and resilient Iowans had other ideas. With resolute strength and determination, they dug deep within themselves to find amazing courage and unwavering faith, hope, and trust.

In each other, they found additional strength and power, and a steadfast resolve to fight back against the greatest of obstacles. Alone, our fragile threads of life can wear thin. Our very existence is threatened. Woven together, our personal fabric transforms into a lifeline of astonishing solidarity. Iowans proved this time and time again as the recovery efforts progressed.

Throughout the agonizing ordeal, thousands of volunteers turned out to help perfect strangers. Natalie Meester and her four-year-old daughter, Anna, of Grundy Center, gave of their time to help in Parkersburg. Numerous national organizations, like Samaritan's Purse and the Billy Graham Rapid Response Team, came to Iowa to aid in the efforts. Prairie Lakes Church of Cedar Falls hosted the Samaritan's Purse volunteers, as congregation volunteers helped them clean up more than 200 flooded homes in the North Cedar area.

In this incredibly historic and important book, meteorologist and author Terry Swails, along with his wife, Carolyn Wettstone, detail, in words and photographs, Iowa's un-natural disasters of 2008. Carolyn and Terry spent weeks interviewing Iowans and personally surveying the astonishing damage. Their efforts in this book will serve as a lasting reminder and testament to the fantastic way Iowans responded with incredible willpower and sacrifice in the face of this catastrophe.

Throughout the disasters, the role of KWWL-TV was, as always, to provide accurate and useful information, as quickly as possible, on the air and online. When the floodwaters wiped out our Cedar Rapids newsroom on the first flood of the Alliant Tower, we brought in reporters and equipment from our sister television stations, owned by our parent company, Quincy Newspapers, Inc. We, too, learned valuable lessons about the strength and power of teamwork and cooperation.

Throughout the course of life, we will face unsettling circumstances far beyond our control. In the face of unimaginable loss, Iowans took this defining moment in our state's history and, without hesitation, stood up to meet the demanding challenges. It's clear Iowa's miracle spirit of character and compassion is very much alive and well. KWWL-TV is proud to salute these heroic efforts.

Ron Steele
KWWL-TV
Senior News Anchor/Reporter

ABOUT THE AUTHORS

Terry Swails is a meteorologist, with thirty-one years of television experience in eastern Iowa. His forecasts have been seen on KWWL-TV in Waterloo, KDUB-TV in Dubuque, and, most recently, at KWQC-TV in Davenport, Iowa. Swails has also published another book titled *Superstorms: Extreme Weather in the Heart of the Heartland*. He is a lifelong resident of Iowa, married to co-author Carolyn S. Wettstone.

Carolyn S. Wettstone is a former television news anchor and award-winning investigative reporter who has worked at KWQC-TV in Davenport, KCCI-TV in Des Moines, and WHBF-TV in Rock Island, Illinois. She is currently a writer and a documentary producer. Her first feature-length documentary, *Sheltering Kevin*, was completed in 2006 and presented at the IMAX Theater in Davenport.

Swails and Wettstone have a daughter, Eden Malone, who was born in 2004.

For more information on this book and other weather-related issues and products, please visit www.terryswails.com.